6·11·79

The Natural History of the Whale

In the past 50 years our knowledge of
whales has increased dramatically – along
with our awareness that these magnificent
animals face a growing threat of extinction.
In this book L. Harrison Matthews has
assembled the most complete and current
picture yet of the natural history of whales:
their physiology and behavior, their
extraordinary relationship to the undersea
environment, and their chances for
survival.

Dr. Harrison Matthews' introductory
chapters sketch the history of man's
interaction with whales and, with the aid of
many drawings and photographs, portray
the basic characteristics of these strange
creatures. He describes each known species
of whale – including the blue whale, the
largest animal ever to live on Earth. Dr.
Harrison Matthews employs the latest
scientific findings to give a fascinating
picture of how whales live. He examines in
detail their feeding habits, their growth and
breeding, the dynamics of their unparallelled
swimming and diving ability, their
migrations, and the social relationships
within schools of whales. He discusses the
dangers whales face from diseases, parasites,
and their chief enemy – man.

Dispelling many of the falsehoods and
half-truths still in circulation, Dr. Harrison
Matthews helps open the way for a clearer
understanding of the largest animals in the
world. His book offers compelling insight
into the lives of these remarkable and still
highly mysterious creatures.

The Natural History of the Whale

L. Harrison Matthews

Columbia University Press
New York

Published in 1978 in Great Britain by
Weidenfeld and Nicolson
and in the United States of America by
Columbia University Press

Printed in the United States of America
10 9 8 7 6 5 4 3 2

Library of Congress Cataloging in Publication Data

Harrison Matthews, Leonard.
   The natural history of the whale.

   Bibliography: p.
   Includes index.
   1. Cetacea.   I. Title.
QL737.C4M315 1978   599'.5   78–2328
ISBN 0–231–04588–3

'Cetological literature is full of poorly supported conjecture'
Carleton Ray and William Schevill

Contents

Plates

Figures

Acknowledgements

The author thanks the Oxford University Press for permission to reproduce the passages from the *Historia Animalium* translated by D'Arcy Wentworth Thompson from *The Oxford Translation of Aristotle* edited by W. D. Ross, Vol 4, 1910, quoted in Chapter 1. He is also grateful to Dr Michael H. Thurston of the Institute of Oceanographic Sciences for his kind help in elucidating Scoresby's names for planktonic Crustacea, to Mrs Frank S. Essapian of Miami for permission to use the photograph of the playful dolphin taken by her late husband, and to Dr James G. Mead of the Smithsonian Institution, United States National Museum, Washington for his helpful comments on the manuscript. The author also thanks Mrs Elizabeth Sutton for the figures, which were drawn especially for the text.

Chapter 1

Cetology from its beginnings to 1945

The scientific discipline of cetology, the study of whales, though rooted in remote antiquity, has grown into a flourishing tree with many branches only during the last two hundred years. It has produced a huge crop of fruit, particularly in the last half century, in the shape of scientific studies on all facets of the natural history of whales.

The naturalists of former times were interested in whales mainly from an academic point of view; they investigated the anatomy and classification of the animals but had little opportunity for observing their physiology or any other aspect of their natural history as living creatures. The great spurt in cetology since about 1920 began when the modern fishery for the great whales appeared to be over-exploiting the whale stocks of the world, so that a source of great commercial profit was being destroyed. Government-sponsored researches, started to gather knowledge of the general biology of whales needed for framing legislation to preserve the whaling industry, led zoologists to an interest in cetology that resulted in far wider investigations than those concerned only with practical problems of commercial exploitation.

The enormous size of the great whales has always been a source of wonder to mankind whenever a casual stranding has given it the chance of examining one at close quarters. Our ancestors of the Stone Age who happened to live near the sea coast were no doubt appropriately impressed but equally overjoyed at the fortuitous provision of so large a source of food and other useful things. Excavations at Skara Brae, a Pictish village in Orkney have shown that some of the inhabitants used the ribs and jaw-bones of whales as rafters in the roofs of their stone-built huts, and the smaller bones for making bowls and other domestic utensils. Although these people lived in the Bronze Age they knew nothing of metals, and their culture was a relic of the Stone Age, cut off from the rest of the world on a remote island. Their main diet was beef, mutton and limpets but they evidently made good use of any whale that happened to be stranded on the shores of their island. Their stone huts were buried nearly to the roofs in the rubbish thrown out during a long occupation, so that in effect they lived down burrows in their own midden [35]. Kitchen middens of great size, up to a quarter of a mile long, 50 feet wide and 20 high, were accumulated on the Danish coast by Neolithic people who lived mainly on shellfish. Among the shells, however, the bones of other animals, including those of whales, show that a chance stranding was no doubt a welcome source of food.

We can never know what ideas early man may have had about the nature of whales, but long before the Skara Brae people founded their squalid settlement the invention of writing by more advanced people elsewhere permits us to know what the ancient civilized world thought of whales and other animals. The earliest writings on animal life that have been preserved to us are those of the Greek philosopher Aristotle, who lived from 384 to 322 BC. Although Aristotle uncritically included in his books many strange tales gathered from travellers, hunters, fishermen and others, tales that are manifestly absurd and contradictory to common observation, he undoubtedly closely studied and dissected many kinds of animal. It is possible that some of the nonsense has been inserted by scribes who garbled his text when copying it. Be that as it may, Aristotle's accounts of whales show that he understood that they are mammals. In his book *On the Parts of Animals* he says 'Some of the water animals such as the whale, the dolphin and all the spouting Cetacea' possess lungs and breathe air [6]. He regards them as intermediate between land animals and the truly aquatic animals such as fishes, and points out that they have no gills but have a blow-hole because they have lungs. He then adds that they cannot prevent sea water getting into the mouth because they feed below the surface, and therefore they have a blow-hole to rid themselves of the unwanted water. He correctly says that the blow-hole is in front of the brain because were it behind it would separate the brain from the spine. 'They are, in a way, land animals as well as water animals; they inhale the air, like land animals, but they have no feet and they get their food from the water as water animals do.'

In the *Historia Animalium* Aristotle tells us much more about the Cetacea. The following passages were translated by the late Professor Sir D'Arcy Thompson who was not only a leading zoologist but also an accomplished classical scholar [7].

The dolphin, the whale and all the rest of the Cetacea, all, that is to say, that are provided with a blow-hole instead of gills, are viviparous. That is to say no one of all these fishes is ever seen to be supplied with eggs, but directly with an embryo from whose differentiation comes the fish, just as in the case of mankind and the viviparous quadrupeds.

The dolphin bears one at a time generally but occasionally two. The whale bears one or at most two, generally two. The porpoise in this respect resembles the dolphin . . . it differs, however, from the dolphin in being less in size and broader in the back; its colour is leaden-black . . .

All creatures that have a blow-hole respire and inspire, for they are provided with lungs. The dolphin has been seen asleep with his nose above water, and when asleep he snores.

The dolphin and the porpoise are provided with milk, and suckle their young. They also take [?carry] their young, when small, inside them. The young of the dolphin grows rapidly, being full grown at ten years of age. Its period of gestation is ten months. It brings forth its young in summer, and never at any other season . . . Its young accompany it for a considerable period; and, in fact, the creature is remarkable for the strength of its parental affection. It lives for many years; some are known to have lived for more than twenty-five, and some for thirty years; the fact is fishermen nick their tails sometimes and set them adrift again, and by this expedient their ages are ascertained.

Over 2,250 years were to pass before more reliable ways of marking whales were invented by modern zoologists for learning about the ages and migrations of whales.

Among the sea-fishes many stories are told about the dolphin, indicative of his gentle and kindly nature, and of manifestations of passionate attachment to boys, in and about Tarentum, Caria, and other places. The story goes that, after a dolphin had been caught and wounded off the coast of Caria, a shoal of dolphins came into the harbour and stopped there until the fishermen let his captive go free; whereupon the shoal departed. A shoal of young dolphins is always, by way of protection, followed by a large one. On one occasion a shoal of dolphins, large and small, was seen, and two dolphins at a little distance appeared swimming underneath a little dead dolphin when it was sinking, and supporting it on their backs, trying out of compassion to prevent its being devoured by some predaceous fish. Incredible stories are told regarding the rapidity of movement of this creature. It appears to be the fleetest of all animals, marine and terrestrial, and it can leap over the masts of large vessels. The speed is chiefly manifested when they are pursuing a fish for food; then, if the fish endeavours to escape, they pursue him in their ravenous hunger down to deep waters; but, when the necessary return swim is getting too long, they hold in their breath, as though calculating the length of it, and then draw

3

themselves together for an effort and shoot up like arrows, trying to make the long ascent rapidly in order to breathe, and in the effort they spring right over a ship's mast if a ship be in the vicinity. This same phenomenon is observed in divers, when they have plunged into deep water; that is, they pull themselves together and rise with a speed proportional to their strength. Dolphins live together in pairs, male and female. It is not known for what reason they run themselves aground on dry land; at all events, it is said they do so at times, and for no obvious reason.

In making his distinction between fishes and cetaceans Aristotle also states, 'Thus the dolphin is directly viviparous, and accordingly we find it furnished with two breasts, not situated high up, but in the neighbourhood of the genitals. And this creature is not provided, like quadrupeds, with visible teats, but has two vents, one on each flank, from which the milk flows; and its young have to follow after it to get suckled, and this phenomenon has actually been witnessed.'

As we shall see in later chapters of this book, most of Aristotle's observations are perfectly correct and show that he had himself examined, and had probably dissected, specimens of Cetacea. The same cannot be said of the natural historians, few as they were, who succeeded him during the following nineteen hundred years. His works were held in such esteem that others were content to accept them as the final truth and preferred to copy them rather than to make original observations of their own. Aelian, who flourished about AD 140 was such an author. Although he was a Roman he wrote his seventeen books of the history of animals in Greek, taking most of his information from Aristotle and embellishing it with legend and travellers' tales. It was, however, Pliny Secundus – Pliny the Elder (AD 24–79) – who wrote the most famous treatise on natural history in Roman classical times. He was a reader, not an observer, so that his works are not full of original information but are an encyclopaedic compilation of things recorded by others. Only his *Historia Animalium*, in thirty-seven books, has survived from his immense output of writings [156].

Chapter VIII of Book IX 'Of Dolphins' begins with a long passage taken straight from Aristotle, but in speaking of the position of the mouth and the way of feeding Pliny muddles what Aristotle said of sharks with his account of dolphins. He repeats the story of the wounded dolphin and its irate companions, and adds many more yarns about dolphins getting friendly with people, including one of a dolphin that daily carried a boy riding on its back across the Bay of Naples to school and brought him back home in the evening. Another, near the city of Hippo on the north African coast, would feed from a man's hand and 'suffer himself gently to be handled, play with them that swom and bathed in the sea, and carrie on his backe whosoever would get upon it'. Pliny adds, 'But there is no end of examples of this kind: for the Amphilochians and Tarentines testifie as much, as touching dolphins that have been enamoured of little boies: which induceth me the rather to beleeve the tale that goeth of Arion.' In Chapter IX, 'Of Porpuisses' Pliny merely says that 'The Porpuisses, which the Latines call *Tursiones*, are made like Dolphins: howbeit they differ, in that they have a more sad and heavie countenence: for they are nothing so gamesome, playfull, and wanton, as be the Dolphins: but especially they are snouted like dogges when they snarle. grin, and are readie to do a shrewd turne.'

Of the great whales Pliny, like Aristotle, has less to say. He mentions the Right whale, *balaena*, and the Sperm whale, *physeter*, but had evidently seen neither; he much exaggerates their size, but correctly says they have lungs connected with the blow-hole on top of the head. He refers to Aristotle's discussion of lungs in cetaceans and gills in fishes, but disagrees with his opinion that fishes do not breathe air but respire by their gills. His only personal observation upon whales was on the Killer whale, *orca*; he says he was present when a Killer got embayed in the harbour at Ostia. The Emperor Claudius, who was supervising the building of a new pier and other harbour works, embarked with a cohort of his Praetorian guards in 'many hoies and barkes' to attack the beast with darts and javelins, 'for to show a pleasing sight to the people of Rome'. Pliny was there and evidently enjoyed the fun, for he says that he himself saw one of the boats 'sunke downe right with the abundance of water that this monstrous fish spouted and filled it withail'.

After the works of Pliny were compiled nothing of importance was added to the literature of cetology, or indeed of biology in general, for nearly fifteen hundred years. During the Dark Ages such matters were forgotten in the West, and the torch of learning was kept burning only by the Arabs, who preserved the ancient literature and added to it the works of such medical authors as Avicenna and Avenzoar, so that a rich store awaited discovery by the medieval schoolmen. We can only guess what more might have come down to us if the famous library of Alexandria, collected at enormous expense from all quarters of the known world by the Ptolemies, had not been burnt by order of the

Caliph Omar in AD 642 – the volumes supplied fuel for the four thousand public baths of the city for six months.

The schoolmen, however, were mainly interested in logic, theology and metaphysics, and showed little interest in natural phenomena. Their respect for the written word held them back from any desire for original observation so that they relied entirely and unquestioningly on the works on the ancients, particularly those of Aristotle and Pliny, for any instruction on natural science that they may have desired.

There is only one exception: the *Kongespeil*, the *Speculum Regale* or *Mirror of Royalty*, an account of Iceland written about 1240, which gives a list of, and some information about, the whales of the Icelandic seas. According to Eschricht [53] it is 'The first writing after Aristotle, and the only one of the middle ages, in which cetaceans have been described from personal observation'. It is the first document in which the differences between the Greenland Right whale or Bowhead and the Northern Right whale, Nordkapper or Sletbag, are recognized. Nevertheless, during the following five centuries the two species, *Balaena mysticetus* and *Eubalaena glacialis*, were confused by zoologists, though not by practical whalers.

It was not until the Renaissance and revival of learning, particularly in the late fifteenth and in the sixteenth centuries, that the yoke of tradition was thrown off and original research was reborn. Two treatises on fishes appeared in quick succession in 1551 and 1554–5, written by the French medical men Pierre Belon (1517–64) and Guillaume Rondelet (1507–66), both containing the results of original dissections. Each writer includes the Cetacea in his works, but only the first definitely regards them as fishes. Belon dissected a porpoise and two species of dolphin, and describes their anatomy in some detail, correctly enumerating all their mammalian characters. Then, as Professor F. J. Cole pointed out, [39] with a complete disregard of reason he rejects the evidence he had himself produced, and classifies the Cetacea among the fishes. This is the more surprising because he entitled his book *The Natural History of Strange Sea Fishes, with the accurate picture and description of the Dolphin and several others of its kind* [15]. He spent several years travelling in the Mediterranean countries, working on the fauna of their coasts, and was murdered in Paris in 1564 at the age of forty-seven. Rondelet, on the other hand, entitles his volumes *A Universal History of Water-Creatures* [170] and so avoids the ambiguity of calling all aquatic animals 'fishes'. He dissected a dolphin and describes much of its anatomy – apparently he was the first to recognize the minute external ear-hole. He distinguishes between the air-breathing Cetacea and the other 'fishes' and compares their structure with that of the pig and of man so that they are 'aquatic quadrupeds' rather than true fishes, but he does not make a complete separation. In 1545 Rondelet became Professor of Medicine and Anatomy in the University of Montpelier where he had been a medical student in the 1530s, as had François Rabelais, who took to medicine when he abandoned the Church. In *Gargantua and Pantagruel* [163] Rabelais introduced a playful portrait of his friend Rondelet as the physician Rondibilis.

For a hundred years after the publication of the books written by Belon and Rondelet little original research on cetaceans appeared, although information about whales both large and small is included in many writings. These fall into several categories. When much of the world was still unknown and unexplored, voyages of discovery were frequent – many of them were written up if the explorers returned, thus adding to the large literature of Voyages and Travels. The sighting of whales is often recorded incidentally and sometimes a brief description of them is given, but beyond adding something to the knowledge of the distribution of whales, they added little to cetology. The famous collections of voyages edited by Richard Hakluyt in 1599, *The Principal Navigations, Voyages, Traffiques and Discoveries of the English Nation* [76]; and by Samuel Purchas in 1625, *Hakluytus Posthumus or Purchas his Pilgrimes* [162] bring together many of the early travels. Purchas includes Captain Thomas Edge's account of the expeditions sent by the Muscovy Company to Spitzbergen, then called Greenland, in 1611 and the ten following years, for exploring and whaling. Edge described the different sorts of whales found, the method of hunting them, and of obtaining their products.

Volumes describing various countries generally include accounts of the vegetation and animals of the particular region, so that many of them have something to say of whales and whaling. Few, however, have anything really new to tell us on the subject. For example, the Right Reverend Erich Pontoppidan, Bishop of Bergen, who published his *Natural History of Norway* in 1751, describes several sorts of whales and gives illustrations of some of them but confesses in the English translation of 1755 [157], 'I have never had the opportunity of seeing a Whale except once, at Sognefaeste, and then he only showed his back above the water, which seem'd to be above forty feet long; and immediately he

7

div'd again.' Even Hans Egede, who spent twenty-five years as a missionary in Greenland and must have seen many whales in that time, gives a rather meagre account of whales and whaling in his *Description of Greenland*, published in Danish in 1741 and in English in 1745 [51]. Although he correctly distinguishes the different kinds of whales, his description of the blowing of the Greenland Right whale is obviously derived from the lurid imaginations of others. He was, however, impressed by the size of a whale's penis, to which he gives a special paragraph signalled by a marginal lemma, and says that it is 'a strong Sinew, seven or eight, and sometimes 14 Foot long'. Sir Robert Sibbald's *Scotia Illustrata, sive Prodromus Historiae Naturalis*, published in 1684 [192], gives a short list of whales, and the Reverend William Borlase's *Natural History of Cornwall* of 1758 [23], gives little over a page to the cetaceans, and this is taken entirely from the works of others. On the other hand Borlase does give fairly accurate engravings of a dolphin and a 'Porpesse', both taken from original drawings made by 'the late Reverend Mr Jago of Loo'.

Several books dealing particularly with fisheries and whaling give abundant information about the methods of catching whales and working up their products, but on the whole much less about the anatomy and habits of the animals though they do give more or less accurate descriptions of the external appearance of the different kinds. One of the best, *Spitzbergische oder Groenlandische Reise-Beschreibung*, was written by Friderich Martens of Hamburg, and published in that city in 1675 [115]. It gives good descriptions of the different whales and of whaling and is illustrated with copper engravings claimed to be done from life. The figures of some of the whales were indeed better than anything that had appeared up to that time, and were copied from book to book for over 150 years. The importance of this work was equalled by C. G. Zorgdrager's *Bloeyende Opkomst der Aloude en Hedendaagsche Groenlandsche Visschery*, published at Amsterdam in 1720 [216], which gives a full history of the northern whale fishery, with statistics of the Dutch and German whaling ships and their catches up to the date of writing. J. A. Allen [1] regarded this as by far the most important source of information on the early history of the subject. The figures of whales were unsurpassed for a century, and were often copied. Some sixty years later in 1782 Duhamel de Monceau published at Paris the four volumes of his *Traité Général des Pêches* [49] in which he included whales and whaling. He deals at length with the catching and processing of whales, but the figures he gives of the different kinds of whale that he mentions are copied from others, so that he adds little new knowledge to cetology. Nearly forty years more had passed when another work of this type appeared, Scoresby's *Account of the Arctic Regions* [184], published at Edinburgh in 1820, a work of great importance that marked a turning point in the scientific study of the oceans and their inhabitants – especially the whales.

By far the largest category of books dealing wholly or partly with whales contains the compendia of natural history and general accounts of the animal kingdom, and the classification systems proposed by various naturalists. Few of the former contain any original information about whales for they purport to be no more than compilations and are direct descendants from the works of Pliny. The earlier ones are little more than bestiaries, often put together uncritically and revelling in fabulous monsters, garbled stories founded on fact, and wonders of all kinds, with a bias towards the usefulness or danger to man of the creatures and their products. Conrad von Gesner, MD, of Basel (1516–65), was a writer of immense production; among his many works he published the *Historia Animalium* in four folio volumes at Zurich, his native city, in 1551–8 [68]. For material he ransacked the writings of all the authors from Aristotle onwards that he could consult. His section on fishes, which includes the whales, is taken mainly from Belon and Rondelet, as are the woodcut illustrations; his descriptions and figures of fabulous sea monsters come from Olaus Magnus. Gesner's *Historia* is regarded by many as the beginning of modern zoology; he died of the plague in 1565 in his fiftieth year. Much of Gesner's work was translated and published in English by Edward Topsel in 1607 [201], but as he confined himself to the 'historie of four-footed beastes', he of course excluded the cetaceans.

A long succession of general natural histories has followed until the present day, those of the seventeenth and eighteenth centuries progressively increasing in accuracy as knowledge improved, though some are little more than plagiarisms of their predecessors. Most of them are written by 'cabinet naturalists' who had no personal knowledge of cetaceans, but they generally include the latest information culled from voyages and the publications of the learned societies. The majority of the authors were mainly interested in classification, the subject reduced to order by Linnaeus (1707–78), the Swedish naturalist, physician, Professor of Medicine and Botany at Uppsala, and scientific adviser to his Government. The tenth edition of his *Systema Naturae* [103] is internationally accepted as the starting point of the modern binomial system of nomenclature. The most widely known of the host of encyclopaedic natural histories are the works of de Buffon [29] and of Cuvier [40]. The Comte de Buffon (1707–88), was a rich man who devoted fifty years of his long life to studying and writing about natural history, having been attracted to the subject through his post of Keeper of the Royal Museum. The first part of his *Histoire Naturelle, Générale et Particulière* appeared in 1749, and publication continued until 1804, in forty-four beautifully illustrated quarto volumes. After his death from vesicular calculus – he refused to be cut for the stone, and fifty calculi were found in his bladder *post mortem* – the last eight volumes were completed by B. G. E. de la Ville La Cépède (1756–1825), later Comte de Lacépède. These included his *Histoire Naturelle des Cétacées* which was also published separately in 1804 [100] and subsequently. The work is a compilation, and the illustrations are mostly copies; when Scoresby was in Paris he met La Cépède on 12 April 1824, and wrote in his diary [197] that although he had criticized the work in his own *Arctic Regions* the author 'was nevertheless very polite and friendly and acknowledged that not having seen whales he had taken everything from research'.

Georges Cuvier (1769–1832), created Baron by Louis Philippe in 1831, held posts in the Paris Museum of Natural History, became Professor of Natural History at the Collège de France in 1799, and later Chancellor of the Imperial University. His extensive original researches, unlike the merely literary researches of La Cépède, were made on the comparative anatomy of recent and fossil animals. The most famous of his works is his *Animal Kingdom – Le Règne Animal Distribué d'après son Organisation* – first published in 1817, followed by numerous editions, and translated into several languages. The works of the lesser compilers, though often lengthy and laborious, are generally puny in comparison with those of these giants. There is nothing original in the *General Zoology or Systematic Natural History*, published in fourteen volumes from 1800 to 1826 by Dr George Shaw (1751–1813), and completed after his death by J. F. Stephens (1792–1853) [191]. The articles in encyclopaedias and other works of reference, though often compiled by professed naturalists, seldom contain original matter.

One may well ask whence came the large mass of material used by all the industrious compilers and plagiarists up to the early years of the nineteenth century, for although much of it was vague and inaccurate it must have had a beginning in original observation at some time. Most of the information about the specific appearance of the different kinds of whale came from the writings of travellers, and from accounts generally derived at second hand from people engaged in the whaling industry. Another source of information was provided by the cetaceans accidentally stranded from time to time on the seashore, particularly the larger whales which often became a nine days' wonder and attracted much public attention. In 1598 the body of a large bull Sperm whale was washed up on the sandy beach at Scheveningen in the Netherlands and crowds of Dutchmen flocked out to see it, including the artist who painted the picture of it still preserved in the Rijksmuseum at Leiden. An engraving was made from this picture and no doubt quickly became a popular print, for it has been copied many times, sometimes in mirror-image when the copier was too lazy to reverse the picture as he engraved his plate; for two hundred years it formed the basis of the figures of the Sperm whale even in ostensibly scientific works. It is not very accurate in details of the head, eye, mouth and flipper, but it does correctly show the great size of the penis, the beauties of which a gentleman is pointing out to his girl friend; someone has long ago prudishly 'painted out' this magnificent member in the Rijksmuseum picture.

On the other hand no illustration appears to have been made of the Right whale that blundered up the Thames estuary in 1658, and was killed near Greenwich. John Evelyn (1620–1706) wrote in his famous diary [25] that on 3 June of that year the whale, 'lying now in shallow water encompassed with boats, after a long conflict, it was killed with a harping iron, struck in the head ... and after a horrid groan it ran ashore and died.' He adds that it 'drew an infinite concourse to see it, by water, horse, coach, and on foot, from London and all parts'.

Sir Robert Sibbald, whom we have already met, was inspired to write a volume published at Edinburgh in 1692 [193] by the stranding of a small school of Killer whales on the shore of the Firth of Forth in 1691. He did not himself see the animals, and relied upon the descriptions of others, but the information he received from those who saw them enabled him to give an excellent description, to which he added some remarks on their habits which he derived from fishermen. He went on to describe a large school of Sperm whales that was stranded in Orkney, and a school of Blackfish stranded in the Firth of Forth. But thereafter he gets into deep water over the description of several different 'species' of Sperm whales through the writings of others being inconsistent with the accounts he had from eye-witnesses of a Sperm whale stranded in Orkney in 1687 and another in the Firth of Forth in 1698. This confusion was perpetuated for over a hundred years by copiers from his *Phalainologia Nova: sive Observationes de rarioribus quibusdam Balaenis in Scotiae Littus nuper ejectis*.

Most of the smaller stranded cetaceans went unnoticed, though some might occasionally be trundled to the nearest town and exhibited as curiosities – or eaten. A few did provide material for anatomical study by naturalists. In 1654 Thomas Bartholin (1616–80) published an account of the anatomy of a pregnant female porpoise that he had dissected in the anatomy theatre of the University of Copenhagen in the presence of King Frederick III and other spectators, under the heading 'Historia XXV, Anatome Tursionis' in his *Historiarum Anatomicarum Rariorum Centuria I et II* [11]. He says that the porpoise abounds in fat from which oil for lamps is extracted, and that the flesh takes the place of bacon in the diet of the poor. He remarks that the internal viscera differ little from those of man, with a few exceptions which he goes on to enumerate. Among others he notes that the stomach has several – he says three – chambers, that the kidneys resemble a bunch of grapes, that the singularly elongated larynx reminded him of a goose's head, and that the end of it lay in the posterior nares. The left side of the uterus contained a male foetus which he also dissected and described.

In 1673 Johan Major (1654–93), a German medical man who became Professor of Medicine and Botany at the University of Kiel, gave an account of the anatomy of a porpoise in the *Miscellania Curiosa* of the Academy of Natural Curiosities for the year 1672 published at Leipzig and Frankfurt in 1673 [111]. The animal was caught by fishermen in the Baltic on 17 April 1671; Major immediately bought it for anatomical use, and dissected it publicly at the University. He gives a pretty full account of the anatomy, but makes the mistake of thinking the blow-hole is for ejecting water taken in by the mouth, and wrongly says that there are no cervical vertebrae except the atlas. Nevertheless it is the most thorough description of the internal anatomy of a cetacean to appear up to this date.

In 1669 John Ray, the English naturalist (1627–1705), dissected a young 'Porpus' found dead on the sands of West Chester, and published a very full account of its anatomy in the sixth volume of the *Philosophi-*

cal Transactions [165] in 1671. It is possible that this animal was not a porpoise but one of the dolphins, perhaps the White-beaked or White-sided dolphin, because he mentions the snout, strong like that of a pig, which he thought might be used for rooting out sand eels, for he found the stomach filled with those little fishes. It cannot have been a Common dolphin because it had only forty-eight teeth in each jaw, and that species has about twice as many.

There is, however, no doubt that the animal dissected by Edward Tyson (1651–1708), a London medical man who was also the first English comparative anatomist of note, was a Common porpoise, for his figure of it in the book he published in 1680 entitled *Phocaena, or the Anatomy of a Porpess* [204] accurately depicts it. His account is much fuller than Ray's, and includes the first description of the retia mirabilia in cetaceans which he then discovered, for they had inexplicably been overlooked by all previous workers; he says they 'formed a curious Net work, and afforded a very pleasant sight'. This work is an important contribution to cetology.

Ray's friend Francis Willughby, with whom he travelled through Europe collecting natural history specimens, died in 1672 at the age of thirty-seven, leaving two large unpublished works on ornithology and ichthyology which Ray added to, edited and prepared for printing. The *Icthyographia* [212], dedicated to the President, Samuel Pepys, and the Fellows of the Royal Society, appeared in 1685. Although the volume is a history of fishes the second book, *De Piscibus Cetaceis*, deals with the whales and dolphins, and in it Ray adds some further particulars to his paper of 1669. In Chapter III '*Phocaena . . . Angl.* A Porpesse' he says that about the end of April 1669, when he was in attendance on the Bishop of Chester, a fisherman found the body of a young porpoise on the beach and brought it to the Episcopal Palace; because it was small and easily handled the Bishop bought it and passed it to Ray for dissection and description. Ray repeats his former description with some additions. He says the snout was longer and more pointed than that in Rondelet's figure and was suitable for digging mussels and little fishes out of the sand. He gives the colour as bluish-black on the back fading on the side into white below; there were no spots or longitudinal streaks as described by Major. He finishes by referring to Tyson's work as more accurate and perfect than his own, and mentions ten points that he missed but Tyson made. He also mentions Bartholin's account of a gravid female porpoise. Although he speaks of the long snout the illustration he gives of *Phocaena* does not show it because the figure is a copy of Tyson's. The figure he gives of the Sperm whale, too, is a copy; it shows the Scheveningen whale in all its masculinity, no doubt to the interest and ribald joy of Samuel Pepys, who paid for the engraving of the plate.

In 1678 Caspar Bartholin Jr edited a corrected second edition of his father Thomas's book *De Unicornu Observationes Novae*, published in Amsterdam [12]. This interesting work contains little that is original; it deals with all kinds of one-horned creatures from the conventional unicorn and other mythical monsters to rhinoceroses and the Narwhal. In discussing this creature Bartholin goes deeply into the nature of the 'horn' and from his examination of this whale's skull, complete with tusk, in the collection of his compatriot Ole Worm, concludes correctly that it is in fact a tooth. Worm (1588–1654), whose name is latinized as Olaus Wormius, was a professor at the University of Copenhagen who made a large collection of natural curiosities which he catalogued in a volume entitled *Museum Wormianum*, published at Leiden in 1655 edited by his son Wilhelm Worm [213]. His chapter on whales is mainly a list taken from the *Speculum Regale*, but the description and illustrations of the Narwhal are from a manuscript of D. Thorlacus Scutonius, the Bishop of Iceland. Bartholin quotes at length from Worm, and gives a copy of his figure of the skull. This figure is also exactly copied, without acknowledgement, in Willughby's *Icthyographia* edited by John Ray.

Throughout most of the eighteenth century little original work on cetaceans appeared until towards its end. There are some scattered notes, mainly on whaling, in the publications of the learned academies, as for example the paper on whales and whaling in New England communicated to the Royal Society by the Hon. Paul Dudley, FRS, in 1725 [50]. This period, however, was the great age of the compiled universal natural histories which naturally contained sections on the cetaceans, with information culled from others. The list of authors is long and includes the names of Hill [84], Brisson [26], Gronovius [75], Erxleben [52], Scopoli [183], Duhamel [49], Bonnaterre [21], Daubenton and Desmarest [46], Kerr [99], Blumenbach [20], and many of lesser competence.

In the latter years of the century original work began again to appear, chiefly as the result of the researches of the anatomists. In 1785 Alexander Monro (1733–1817) published at Edinburgh his *Structure and Physiology of Fishes*, illustrated with engravings of dissections, including some parts of a porpoise that he had himself dissected [130]. He was Alexander Monro *secundus*, the son of Alexander *primus* (1697–1767), and father of Alexander *tertius* (1773–1859), all of whom were successively Professors of Anatomy in the University of Edinburgh.

But the real pioneer in cetological anatomy of the times was John Hunter (1728–93). He was Scotch by birth, from Kilbride near Glasgow, but spent his working life in London whither he came at the age of twenty to learn medicine under his brother William, ten years his senior and already well known as a surgeon and teacher of anatomy. John duly became a surgeon and worked for many years at St George's Hospital, where he died of a heart attack after a heated argument about students' fees at a meeting of the Hospital Board on 16 October 1793. Both brothers accumulated vast collections of anatomical specimens, that of John covering not only human anatomy but the comparative anatomy of animals in general. After their deaths William's collection went to the University of Glasgow, but John's was bought by the Government for £15,000 and housed in the Royal College of Surgeons at Lincoln's Inn Fields.

John, among his numerous researches, was particularly interested in the cetaceans; he even went to the expense of sending a surgeon north in a Greenland whaling vessel to gather information for him. He published his collected observations in a paper in Volume 77 of the *Philosophical Transactions of the Royal Society* in 1787 [90]. This contains an account of the anatomy of several species that he had dissected, including the first recorded British specimens of *Tursiops truncatus*, the Bottle-nosed dolphin, captured in the Severn estuary near Berkeley and sent to him by his former pupil and lifelong friend Dr Edward Jenner (1749–1823), the introducer of cow-pox vaccination as a protection against smallpox. He also for the first time elucidated the structure and growth of baleen in a 17 foot Piked whale (Minke whale), *Baleanoptera acutorostrata*, caught on the Dogger Bank. This paper and its engraved illustrations were reproduced in the collected works of Hunter edited in five volumes by James Palmer, senior surgeon to St George's Hospital, in 1837 [147].

At his death John Hunter left an enormous mass of unpublished manuscripts and of what would now be called laboratory notebooks, recording his dissections and his thoughts about comparative anatomy and surgery. His brother-in-law, Everard Home, a disreputable character far different from his sister Anne, Hunter's wife, was one of his executors. Everard Home was nevertheless a rich and fashionable surgeon, and a baronet through being a crony and boon companion in the drunken debaucheries of the Prince Regent, later George IV. He also lectured on anatomy and fancied himself as a man of science, collaborating with Hunter in several papers given to the Royal Society, of which he was elected a Fellow, and publishing books on anatomy and pathology. After Hunter's death he illegally got possession of the notes and unpublished manuscripts, from which he stole the material for a long series of papers in the *Philosophical Transactions* – his total reached 143; and then burned the lot to cover his tracks, setting his chimney on fire in the process. But all was not lost. Hunter's secretary, William Clift (1775–1849), who became Curator of the Hunterian collections at the Royal College of Surgeons, had made copies of most of them, and gave them to his assistant, Richard Owen (1804–92). Owen, who married Clift's daughter in 1835, was later Sir Richard Owen, FRS, Hunterian Professor at the College and finally Superintendent of the natural history collections of the British Museum. He edited and arranged the transcribed Hunterian works, and published them in two volumes in 1861 [142]. They cover the whole field of human anatomy and zoology, but those of particular interest to cetologists are the original materials from which the famous paper on whales in the *Philosophical Trans-actions* was written.

The Reverend William Scoresby Jr, DD, FRS (1789–1857) made his first whaling voyage to the Arctic with his father, William Scoresby Snr, at the age of eleven. At the age of twenty he was master of the whaler *Resolution*; he made his last annual voyage at the age of thirty-three, then went to Cambridge and became a parson. The best known of his works are two books on the Arctic – *An Account of the Arctic Regions with a History and Description of the Northern Whale-fishery*, published in two volumes in 1820 [184] and *A Journal of a Voyage to the Northern Whale-fishery*, published in 1823 [185]. The first deals with polar exploration, physical oceanography, meteorology and the fauna of the Arctic, and gives a history and very detailed account of the whale fishery and its methods. Scoresby was the first scientific man to know anything about the larger whales as living animals – he learnt much during his twenty years of hunting and cutting them in, in particular the anatomy, habits and food of the now rare Greenland Right whale, his main quarry. The second book describes a survey of five hundred miles of the east coast of Greenland, with observations on meteorology, magnetism and physiography, but includes an illustrated account of the anatomy of the Narwhal and of the Greenland Right whale.

From his earliest voyage Scoresby was fascinated by the shapes of snow crystals, and gives four engraved plates with ninety-six figures of them in his first work. In his early days in his father's ships he used to astonish the sailors by lighting their pipes with the sun's rays focused through a lens of ice. The experiment of producing fire with a burning 'glass' of ice had, however, been made and recorded by J. D. Major nearly a hundred and fifty years before in 1672 [112]. Scoresby's work had not been appreciated at its true worth until the revival of interest in cetology in the first half of the present century, though his biography written by his nephew appeared in 1861 [186]; a recent biography [197] adds much of interest because the authors, T. and C. Stamp, have been able to consult the Scoresby correspondence and other papers preserved in the museum of Whitby, his home port.

With the works of Scoresby we come to the nineteenth century, in which the science of cetology made great progress, though mainly in systematics, taxonomy and anatomy, rather than biology. An ever increasing number of papers appeared in the growing family of journals published by the learned societies but they are too many to be discussed in detail here, though some of the more outstanding researchers and their results must be mentioned.

The numerous papers by Everard Home published in the *Philosophical Transactions* during the first quarter of the century, many on cetaceans, and taken without acknowledgement from Hunter's work, have already been mentioned. In 1823 an enlarged edition of G. Cuvier's *Ossements Fossils* [41] appeared with much new information on the skeleton of both fossil and modern cetaceans, mostly reprinted from the *Mémoires* of the Paris Museum. From the 1830s onwards a long series of papers came from the pen of Owen, most of them palaeontological but some on the systematics and anatomy of modern cetaceans, notably a beautifully illustrated paper in the sixth volume of the *Transactions* of the Zoological Society [144]. J. E. Gray, FRS (1800–75), Keeper of Zoology at the British Museum, was equally industrious, publishing an enormous amount of work on many aspects of zoology, including the Cetacea. He described and named a great number of false species from cetacean skulls and other skeletal fragments sent to the Museum – but what else could the poor wretch do when confronted with a specimen that differed from anything previously known? His employers required him to conserve appropriately everything that came into their trust, the extent of intra-specific and age-dependent variation was unknown, so Gray's labours resulted in a multiplication of synonymies and later confusion.

Shortly after the middle of the century several zoologists in Scandinavia published important papers on various cetaceans: D. F. Eschricht (1798–1863), Professor of Physiology in the University of Copenhagen, W. Lilljeborg (1816–1903), Professor of Zoology in the University of Uppsala, and J. Reinhardt (1816–82), Professor of Zoology at the Zoological Museum of Copenhagen University. They were considered to be so valuable that Professor W. H. Flower, FRS (1831–99), Conservator of the Museum of the Royal College of Surgeons, and later Sir William Flower, Director of the British Museum (Natural History), translated them for publication in English by the Ray Society in 1866 [57]. Flower himself wrote many papers on cetaceans, as did F. E. Beddard, FRS, for over thirty years Prosector to the Zoological Society of London until he retired in 1915. In Scotland Sir William Turner, FRS, Professor of Anatomy at Edinburgh University, and later Principal, published numerous memoirs on the various Cetacea he dissected. Many other zoologists and anatomists contributed papers from time to time to the publications of the scientific societies when they had the opportunity of observing cetaceans alive or dead, but of books devoted solely to the whales and dolphins there were remarkably few.

It is not always easy to distinguish them from the general natural histories, encyclopaedias, faunas, catalogues of the contents of museums, or reports of the results of exploring expeditions, for sometimes the matter about the cetaceans was sufficient to make a whole volume in a series. The volume on Cetacea, which is no more than a compilation, in the general natural history [102] published in 1828 by the French naturalist R. P. Lesson (1794–1849), is an example, as is that by an unnamed author on *The Ordinary Cetacea or Whales* [2] published in 1837 in the Naturalists' Library conducted by Sir William Jardine (1800–74). In 1834 Henry Dewhurst, Surgeon-Accoucheur, a self-advertising medical quack, published a book on the natural history of the Cetacea and of the Arctic regions [47]; it is mostly a compilation in imitation of Scoresby's book, though it does contain some original observations made in 1824 when he went on a voyage to the Greenland seas as ship's surgeon in the whale ship *Neptune* of London. On the

18

other hand in 1820 an edition of the original works on Cetacea by the eighteenth-century Dutch naturalist Peter Camper (1722–89), was published in Paris by his son Adrien-Gilles Camper [33]. His researches on cetaceans produced much original information about the skeleton and soft parts of the animals he examined.

In 1835 Thomas Beale, who had served as surgeon to the *Kent* and *Sarah and Elizabeth* southseamen on a whaling voyage from 1830 to 1833, published *The Natural History of the Sperm Whale* with an account of the Sperm whale fishery. Although he had first-hand knowledge of his subject his natural history consists largely of quotations from others. Nevertheless the book was widely sold so that in 1839 he published a second edition [13] to which he added 'A Sketch of a South-Sea Whaling Voyage' in the form of an abbreviated log of his wanderings in search of whales in the South Pacific. One of the authors he quotes extensively in his account of the anatomy of the Sperm whale is F. D. Bennett, who had likewise made a south sea whaling voyage immediately after Beale's, from 1833 to 1836, and had made several original communications on the Sperm whale to the Zoological Society of London, which were printed in its *Proceedings* [16, 17]. He later brought his observations together in his *Narrative of a Whaling Voyage round the Globe* published in 1840 [18]. Bennett's work was really original, and is a valuable contribution to cetology. So too is the book on the Cetacea published three years before, in 1837, by W. Rapp, Professor of Anatomy at the University of Tübingen [164]. The first part of it deals with systematics, but the second with anatomy, based upon the original observations of the author.

There appears to be no original work in the none the less valuable and critical compilation on the Cetacea [43] published in 1836 by Frédéric Cuvier (1773–1838), the less known younger brother of Georges, Baron Cuvier. The anatomical part of this work was largely reprinted in Todd's *Cyclopaedia of Anatomy* [42] in the article 'Cetacea' in the first volume, published in 1835–6, and although it has additions from Hunter's work it is signed with his name. Flower's translation of three papers by the Danish cetologists has already been mentioned; it was published as a single volume by the Ray Society in 1866. Both Gray [72] and Flower [60] compiled catalogues of the cetacean material in the British Museum, which were issued as separate volumes in 1866 and 1885 respectively. On a more modest scale Thomas Southwell (1831–1909), of Norwich, published a popular but accurate compilation on the seals and whales of the British seas [196] in 1881.

In the 1880s two important works on cetology were written by Belgian zoologists: a classic on the skeletal anatomy of recent and fossil cetaceans by van Beneden and Gervais [206] published in 1880, and a second, hardly less important, by van Beneden on the natural history of the cetaceans of European seas [205], which was first published in the *Mémoires* of the Royal Academy of Sciences of Belgium in 1885 and republished as a separate volume in 1889. In the last year of the century F. Beddard, already mentioned, published his semi-popular *Book of Whales* [14], in which he competently brought together the current knowledge of the Cetacea gathered from the works of others and his own original anatomical researches.

In the United States of America, although many papers on cetaceans, notably by Professor E. D. Cope and Dr W. H. Dall, appeared in various journals, the outstanding book giving some account of whales as living animals was written by the whaling captain C. M. Scammon and published in 1874 [175]. The book has become a classic, being filled with original observations made by Scammon or gathered from his brother whaling captains. The most valuable part is his description of the Gray whale and its natural history, and of hunting it in the coastal lagoons of Lower California, one of which was known even in his time as Scammon's Lagoon, now sunk to the ignominy of a tourist attraction.

The Whaleman's Adventures in the Southern Ocean by the Reverend Henry T. Cheever, a 'pious and observant American clergyman', adds little new to cetology, though it is a readable narrative of events [34]. Cheever, who made a voyage lasting over eight months 'on account of his health' in the whale ship *Commodore Preeble* homewards from his missionary labours in the south Pacific, published this account of his adventures, and of whales and whaling in general, in the USA, but the English edition published in 1850 was edited and modified by William Scoresby, then living in retirement at Torquay. In 1878 Alexander Starbuck published a comprehensive 'History of the American Whale Fishery' up to 1876, with valuable tables listing all ships, their voyages and catches [198]. This was followed in 1881 by J. A. Allen's comprehensive bibliography of works on cetaceans and sirenians [3], and soon after by an account of the whales and porpoises by G. B. Goode in the report he edited on the fisheries of the United States [70].

Early in the twentieth century cetology received a great stimulus by the publication in 1904 of Volume XXXIII of the *Smithsonian Con-*

tributions to Knowledge, a weighty monograph on the whalebone whales
of the western north Atlantic written by Dr F. W. True, Head Curator
of the Department of Biology in the US National Museum [203]. He
owed his opportunity for making this detailed study to the establishment of a shore whaling station run on Norwegian lines near St
John's, Newfoundland, where he was able to examine many specimens
of several species of whales. He also brought together all the relevant
information on the whalebone whales written by previous authors to
resolve the confusion arising from the many synonymous scientific
names and to establish the characters of the various species.

A similar study, though on a smaller scale, of southern whalebone
whales was made by G. E. H. Barrett-Hamilton in 1913 at the whaling
stations in South Georgia. A Government Committee on Whaling and
the Protection of Whales had been set up to recommend regulations to
protect whales from over-exploitation by the then young Antarctic
whale fishery, and Barrett-Hamilton was sent to gather information on
the biology of the whales as a preliminary to framing legislation.

The scientific part of the report, written by M. A. C. Hinton after
Barrett-Hamilton's death in 1914 [85], concentrates on the breeding biology of whales, and the discussion emphasizes the gross over-fishing of the
Humpback whale which as early as 1913 was already threatened with
extermination. The reconstituted Committee organized the *Discovery*
Expedition in 1924 to gather more information on the biology of whales
and Antarctic ecology, an object most successfully attained during the
ensuing twenty-five years of the *Discovery* Investigations. As all the
world knows, the immense expenditure of time, effort and money was
futile, for the information that should have guided the whaling industry
was constantly disregarded, the populations of whales were severely
reduced and the whaling industry, once profitable in material and
money, was ruined. Only at the eleventh hour has the International
Whaling Commission put a lock on the stable door after most of the
horses have gone, and even then it has given a duplicate key to Russia,
Japan and some others, and is unable to prevent them from using it. On
the other hand the scientific value of the Investigations to cetology has
been immense, so that our knowledge of the biology of the larger
Cetacea is now as great as that for any other order of wild animals.

The *Discovery* Investigations continued until, and were resumed
after, the war of 1939–45, after which the modern phase of cetological
research began. In the meantime, apart from researches based on the

whaling industry and its effects on the whale stocks, the growing literature of whale biology was augmented by the scheme started by Sir Sidney Harmer, FRS, then Director of the British Museum (Natural History), who secured the co-operation of the Coastguard, Customs, and the Receiver of Wreck, by which all Cetacea stranded on the British coasts should be reported to the Museum. Over the years this arrangement, started in 1912, has resulted in the accumulation of much knowledge and museum material for cetology, reports on which continue to be published from time to time [78].

The study of cetology grew rapidly after the end of war in 1945, owing partly to the continued anxiety about the decline of whale stocks through excessive hunting, but more to the new techniques in oceanography developed from wartime inventions, to the large funds given by Governments for scientific research, and to the establishment of huge marine aquaria where, for the first time, the smaller Cetacea could be studied alive in captivity, and their biology scientifically investigated. The results of all this work are far too large for any attempt to follow them chronologically here, but much of the essence of them is distilled and presented in the pages that follow.

Chapter 2

The cetacean diversity

Although the great whales are truly great, they form only a minority of the many species of the Cetacea. By far the largest number of the animals making up the order are the comparatively small dolphins and porpoises. As with some other orders of the mammalia we have no certain knowledge of their origin, for the earliest-known fossils from the Eocene are already unmistakably whales, and we can only guess at their evolutionary history by inference.

The architecture of the body of all cetaceans is that of a land mammal that has taken to a completely aquatic way of life, so that in spite of many unique modifications in detail, their structure and organs are homologous with those of all other mammals, aquatic, terrestrial or aerial. Thus they breathe air by lungs, are warm-blooded and give birth to living young that are suckled on milk secreted by the mammary glands of the mother. All this was known to Aristotle so that it is surprising that the Cetacea were confused with fishes for so long even by the literate, in view of the veneration in which his works were held. Although the various kinds of Cetacea differ widely among themselves their outward appearance has several features in common: the naked, fusiform, torpedo-shaped or fish-like body, the fore-limbs modified as paddles or 'flippers', the absence of hind limbs, the expansion of the end of the tail as horizontal flukes, and the nostrils opening as single or double blow-holes far behind the end of the snout.

The position of the nostrils on top of the head is peculiar to the Cetacea – no other mammals have similar blow-holes. The advantage of this arrangement to an aquatic mammal is obvious, but the modifications of the skull to produce it in the course of evolution are complex. The process by which this has come about can be pictured by imagining the bones of the skull of a typical mammal, with nostrils at the tip of the snout, as made of plastic clay being moulded by a sculptor's

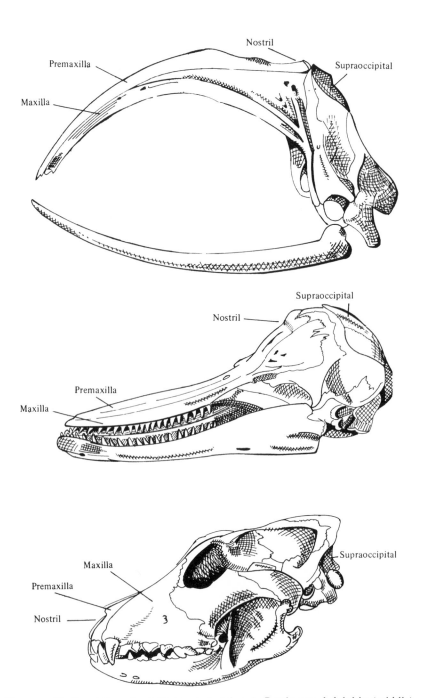

Figure 1. Skulls of Greenland Right whale (above), Bottle-nosed dolphin (middle), and dog (below), to show the 'telescoping' of the cetacean skull compared with that of a land mammal.

hands. The upper jaw-bones, the maxillae, are pulled forward to make a beak or rostrum, and at the same time the premaxillae are extended backwards from the tip of the rostrum to push the nostrils back towards the frontal bone. As this happens the nasal bones, which roof the nasal passage, are pushed back towards the frontal, and greatly reduced in size, so that in some species they are represented only by a small tubercle on that bone. While these bones of the fore part of the skull are extending backwards, the supraoccipital bone at the back of the skull is drawn upwards and forwards to cover the brain and form the posterior part of the roof of the brain case. So extensive is its proliferation that it pushes the parietals, which in most mammals form the sides and roof of the cranium, away to right and left so that they do not make a suture in the mid-line. The frontal is thereby much reduced in size so that it takes the form of a fillet on the forehead, expanded laterally over the orbits.

As these changes are made numerous modifications of the other bones of the skull are necessary. The ethmoid thus forms the greater part of the front of the cranium, supported and partly overlain by the unusually large expanded pterygoids, which, with the vomer, extend forwards, largely usurping the place of the palatine. At the same time the jugal arch is much reduced, often to a slender bar below the orbit, and in some species is completely lost; but the squamosal remains relatively large. The ear bones, the tympanic and periotic, are only loosely articulated with the rest of the skull.

The combined result of these modifications is that the brain case is short but high and wide, the nasal passage lies approximately vertically in front of it, and a comparatively long rostrum extends approximately horizontally before. This basic cetacean pattern is modified in various directions in the different suborders and families.

The order Cetacea is divided into three suborders; the Archaeoceti, all of which are extinct and known only from fossil remains, the Odontoceti or toothed whales, containing the majority of the living species forming the order; and the Mysticeti, the whalebone or baleen whales.

The Archaeoceti, as their name implies, are the oldest group and consequently the most primitive in their structure. They flourished during the Eocene epoch but most of them were extinct before the end of the Oligocene, only a few small species surviving into the early Miocene. The most primitive species, such as *Protocetus*, were elongated aquatic animals with reduced hind limbs and long snouts. The

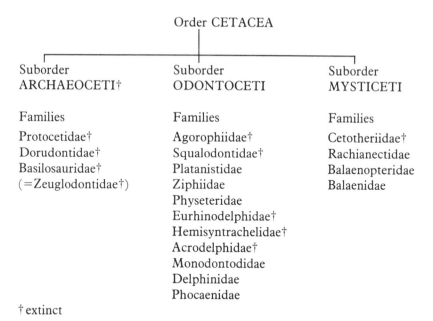

Figure 2. The classification of the Cetacea.

bones of the skull are all arranged in the typical mammalian pattern without the distortion or 'telescoping' shown by modern Cetacea. The only sign of events to come is the position of the nostrils on the top of the snout or rostrum some way back from the tip. The teeth, too, are completely different from those of modern whales for they are heterodont, and distinguishable as incisors, canines and grinders, the last sectorial rather than molar, probably in correlation with a diet of fish. The dentition and general form of the skull show that the early Cetacea were probably descended from a common stock with the primitive Creodonta, which also gave rise to the Artiodactyla as well as the Carnivora.

The most striking and characteristic family of the Archaeoceti is the Basilosauridae or Zeuglodontidae, members of which were abundant in the seas of the upper Eocene. They were large animals, some species reaching a length of 70 feet, and having skulls 5 feet long. Unlike the modern whales they were not fusiform in body shape but snake- or eel-like; elongated and cylindrical. The neck was short but the rest of the vertebral column very long; the fore-limbs were short paddles and the hind limbs had been lost.

The form of the teeth suggests that these creatures fed upon fish, but the snaky shape of the body suggests that they may have differed strikingly from the rest of the Cetacea in the shape of their tail flukes, if indeed they had any. The snake- or eel-like body leads to speculation on their way of swimming. Did they swim by vertical oscillations of horizontal tail flukes, or did they swim by lateral flexures of the whole body, as in the animals they resemble, albeit on an enormous scale? The vertebrae each bear dorsal and lateral processes which are comparatively small so that there was no possibility of their articulating closely with each other or being tightly bound together with ligaments. This gave the vertebral column a degree of flexibility and freedom of movement that is denied to the modern Cetacea which have much more massive processes articulating to form a more rigid spine. It is therefore possible that the basilosaurs lived like eels in comparatively shallow water, snaking their way over the bottom to snap up the fish and Crustacea of the benthic fauna. If this were so, they probably did not have horizontal tail flukes, nor indeed flukes of any sort: the hind end of the body and the tail were no doubt laterally compressed, as is the tail-stock of all modern cetaceans, and this would give all the purchase on the water needed in eel-like swimming.

The long snake-like form of the zeuglodonts misled the palaeontologists who examined the first specimens discovered into thinking they were dealing with the remains of an extinct reptile, hence the name *Basilosaurus* bestowed upon them in 1834 by the American geologist Harlan but altered in 1839 to *Zeuglodon* by the English anatomist Owen, who based his name on the character of the molar teeth, which have two roots joined by a blade-like crown.

In the second quarter of the nineteenth century an American fossil hunter, Dr Albert C. Koch, collected a large number of fossil vertebrae and other bones of archaeocetes from the Eocene beds in Alabama and South Carolina, where they are abundant. Although the vertebrae came from several individuals Dr Koch strung them together to make the skeleton of a supposed animal 114 feet long, which he exhibited as the skeleton of a sea-serpent in a hall on Broadway, New York, in 1845. Zoologists were quick to denounce the imposture but nevertheless the public, as always eager for marvels, flocked to see it so that the self-styled doctor made a good thing out of his humbug. He even went so far as to give the creature a scientific name, *Hydrarchos sillimani*, in honour of Professor Benjamin Silliman, a genuine scientist who for many years

27

edited the *American Journal of Science and Arts*, also known as *Silliman's Journal of Science*. The doctor, whose own name is not inappropriate, seems to have had a cynical sense of humour in selecting the name of the well-known professor for his fabrication. Now we have something even sillier – *Nessiteras rhombopteryx*, P. Scott 1975, the non-animal of Loch Ness.

Among the Archaeoceti the Protocetidae may represent the stock that gave rise to both groups of living whales, whereas the Basilosauridae, although having skulls and dentition that were primitive as compared with those of modern whales, were highly evolved and specialized in their post-cranial structure. They can in no way be regarded as the progenitors of the order Cetacea; they are no more than an aberrant branch from the common stock of origin.

Most of the members of the Odontoceti are comparatively small porpoises and dolphins, though some, such as the Beaked whales and the Killer whale, reach a length of 30 feet and one, the Sperm whale, reaches 60 feet or more. In most of the odontocetes the jaws are prolonged more or less as a beak-like snout behind which the forehead rises in a rounded curve that is very prominent in some species in which it is called the 'melon': *melon* in colloquial French means a bowler hat. Even in those species that lack a beak the rostral part of the skull is prolonged forwards so that if the flesh is removed the skull shows the same general form as that of the beaked species.

The Odontoceti differ from the other suborders in the arrangement of their nasal passages and blow-holes; they differ indeed from all other mammals in having only a single nostril. The nasal passages are separate at the base of the skull, as is usual, but in their passage to the surface they join close below it to form a single opening or, in extreme cases, one is functionally suppressed leaving the other as the sole breathing tube. The blow-hole is typically a crescentic slit, with the horns directed forward on the summit of the head; it is closed by a fatty and fibrous pad or plug that lies between the horns. When the animal breathes it thus has to open the blow-hole by muscular effort; when it dives the pressure of the water keeps the hole sealed. There are, in addition, various side branches and valvular arrangements inside the blow-hole which will be discussed below when dealing with the respiration of the Cetacea (Chapter 5).

The skull bones of the nasal region, especially the premaxillae and the nasal bones, are often asymmetrical in their size, shape and position.

That is not to say that the skull as a whole is asymmetrical, but that the arrangement of some of its component bones is different on opposite sides. In the Sperm whale, in which the left nasal passage alone serves for breathing, so that there is a single blow-hole of unusual shape on the left side of the snout, the correlation of the asymmetrical skull bones with the asymmetrical form of the nasal passages can be readily appreciated. In other species, however, where the blow-hole passage is less asymmetrical, the universal, though less pronounced, asymmetry of the bones of the upper part of the skull is less easily understood. This point is discussed more fully below.

The Odontoceti always bear teeth in the upper or lower jaws, or both, at least in some stage of their lives. The teeth always have a single root so that they are simple or slightly recurved pegs set in single sockets. They are never differentiated into incisors, canines and molars, but all look much alike. Some species have many teeth in both upper and lower jaws, others have few; some have teeth only in the lower jaw, some when fully adult have none at all. The dentition is monophyodont, that is, there is only one set of teeth, and no succession of a milk dentition by a permanent one.

The dorsal fin is another character that is generally present but is lacking in some species. When present it is never supported by bones inside: like the tail flukes it consists of firm fibrous and fatty tissue.

Turning now from the features that are common to all odontocetes it is convenient to examine those that distinguish the six families of living species into which systematists divide the suborder. First, however, we must glance at one of the families containing only extinct species, which are regarded as the most primitive members of the odontocetes, the Squalodontidae. The animals were much like modern dolphins in general appearance and habits, but take their name from the triangular teeth, resembling those of sharks, that arm the hind part of the jaws; the telescoping of the skull bones, however, was much as in modern odontocetes. The squalodonts first appear in the upper Oligocene, and were abundant during the Miocene, declining towards its end and becoming extinct early in the Pliocene. In the Miocene they were contemporary with a large number of now extinct species of the family Delphinidae, which contains the dolphins and porpoises now living. Most of the Squalodontidae are not regarded as ancestral to the other odontocetes, but as collaterals descended from a common stock, though some of the earliest forms may have been.

The family Platanistidae is thought by some zoologists to contain the nearest living relatives of the squalodonts, though they differ in having simple peg-like teeth. They resemble them, however, in that the temporal fossa of the skull is not roofed by parts of the maxillae as it is in the other living odontocetes. The platanistids are all fresh-water dolphins inhabiting some of the great rivers of Asia and America; only one is also found in the sea. They are not the only river dolphins, however, for some of the members of the family Delphinidae also inhabit fresh waters. They are peculiar-looking creatures: all have a long beak and rather broad short flippers, some species have a low hump-like dorsal fin; in *Platanista* of Indian rivers the blow-hole is a longitudinal slit, but in *Pontoporia* of the River Plate and nearby coasts it is crescentic. In some, too, there is a distinct indication of a neck between head and body owing to the cervical vertebrae being separate and not partly or wholly fused together as in most cetaceans. The eyes are comparatively very small, and at least one species is blind. The life of these interesting creatures is considered in more detail below (Chapter 8).

The family Ziphiidae also contains some very peculiar animals. They are small to medium-sized whales ranging in length from 15 to 30 feet, although one species sometimes reaches over 40 feet. In all of them the

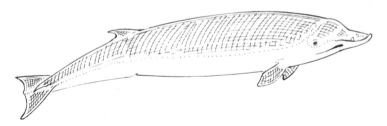

Figure 3. Sowerby's Beaked whale.

jaws form a beak, strongly marked off from the rest of the head in some species, hence their name, 'Beaked' or 'Bottle-nosed' whales. There are two longitudinal grooves on the throat, the dorsal fin is set rather far back on the body, and the flippers are rather small. The most peculiar character of the ziphiids, however, is the dentition. In all except one species the teeth are very reduced in number and entirely absent from the upper jaw. In the lower jaw of the adult males one or two teeth only on each side are generally present; they are comparatively large and sometimes project from the mouth as small tusks. In the young males and the females these teeth generally do not cut the gum in which they are embedded, so that the animals appear to be entirely toothless. The family contains five genera embracing about fifteen species, some of which are found widely throughout the oceans of the world. Nevertheless most of the species are very imperfectly known to science, from specimens occasionally stranded; they are consequently considered rare animals though they must be common enough in parts of the ocean far from the coasts, as is shown by Gosse's experience in 1844. P. H. Gosse, FRS (1810–88), the English naturalist, was on a voyage to Jamaica, and on 22 November when in mid-Atlantic about half-way between the Cape Verde Islands and the northernmost of the Leeward Islands, in latitude 19°N. and from longitude 46° to 48°W. saw a school of Beaked whales which followed and played around his ship for no less than seventeen hours continuously, 'during the whole of this time, the ship had been running before a gallant breeze, and had proceeded nearly 120 English miles'. [71] He could not identify the species of Beaked whale that he saw; it was probably one then unknown to science. As an example of confusion based on the examination of inadequate material one species of Beaked whale, *Mesoplodon hectori*, was said after ninety years to be merely the young of another species *Berardius arnuxii* [121], but further research has now thrown doubt on this opinion.

Only one species of ziphiid can be said to be well known, the common Bottle-nosed whale *Hyperoodon ampullatus* which was formerly the subject of a fishery and is still hunted commercially on a smaller scale. It is a small whale, the males reaching a length of about 30 feet, the females not more than 25. It is common in the north Atlantic and Arctic Oceans – a closely similar species is numerous in the southern hemisphere – and generally occurs in medium-sized schools. The beak of the Bottle-nosed whale is particularly distinct, forming the neck of the 'bottle'. It is most prominent in the old males because the rounded forehead behind it is greatly enlarged; the pad of the melon, unlike that in the dolphins, is supported by wide bony crests developed by the hinder part of the maxillae. The two teeth at the tip of the lower jaw sometimes penetrate the gums in old males, but in females and younger males they do not erupt. In the heyday of the fishery of Bottle-nosed whales the oil they yielded was known as 'Arctic sperm oil'.

Members of the Ziphiidae have been found in all the oceans; those in the genera *Berardius* and *Ziphius*, like *Hyperoodon*, have teeth at the tip of the lower jaw, but the males of all but one of the several species of the

31

genus *Mesoplodon* have a single tooth about half-way along the lower jaw. This tooth or tusk is large in some species, and reaches its greatest length in the bizarre Layard's whale, in which it curves over the upper jaw outside the mouth so that the opening of the jaws is limited. The most peculiar of the Beaked whales was discovered during the 1930s on the coast of New Zealand [140], and has since been found in South American waters. Like others it has two large teeth at the tip of the lower jaw, but it is unique in also having about twenty smaller functional teeth on each side of each jaw. This justifies its being placed in a new genus *Tasmacetus*, although other Beaked whales often have some rudimentary jaw teeth that do not cut the gums.

In coloration the Beaked whales follow a general pattern of dark, often black, on the back shading through grey to light grey or white on the belly, but there is much variation between individuals and according to age. Aged animals appear to lose much of their pigmentation, especially on the head. But two specimens of Cuvier's whale *Ziphius cavirostris*, one of them a young female, the other an adult male, have been seen in which the pattern was reversed, the under parts being dark but the head and fore part of the body and back creamy white [79].

The family Physeteridae contains only three living species, the cosmopolitan Sperm whale, the Pygmy Sperm whale found in the Atlantic, Pacific and Indian Oceans, and the little known Dwarf Sperm whale of tropical waters. The Sperm whale, the largest of the toothed whales, full-grown males reaching a length of 60 feet, is one of the most peculiar looking whales, matched for its bizarre proportions only by the Bowhead among the baleen whales. The enormous barrel-like head accounts for about a third of the animal's length, and behind it the body tapers to the tail flukes; there is no dorsal fin, but several low humps lie aft of amidships on the back. The lower jaw is very narrow and does not reach the end of the snout; it carries some twenty to thirty teeth on each side. The upper jaw is devoid of functional teeth though it often carries a few rudimentary ones in the gum, which is tough and rubbery with a number of sockets into which the lower teeth fit when the jaws are shut. The eye lies close to the angle of the mouth, and behind it lies the large rounded paddle or flipper.

The internal structure of the enormous head is interesting, for the shape of the skull is quite unlike that of the living head. The comparatively small brain case lies at the back, and in front of it a wide shelf of bone runs forward, tapering to a blunt point at the snout. The bones

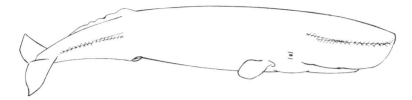

Figure 4. Sperm whale.

of this shelf, which forms the upper jaw, are the premaxillae and the maxillae, with the long vomer in the centre line below. At the rear end the maxillae expand into curved plates that sweep upwards and inwards to join the large supraoccipital that covers the brain case; between them these bones make a curved crest enclosing a chair-shaped bowl at the bottom of which lie the passages for the nostrils, large for the left, small for the right. The bones of each side of the skull contribute unequally to this arrangement so that they are widely asymmetrical, as are the much smaller nasal and frontal bones which take a minor part in the structure. The soft tissues of the head sit in this chair, the 'Neptune's sledge' of the old whalers, and at the front end rise far above the supporting shelf of the upper jaw. The beautiful lithographed plates illustrating Flower's paper of 1869 [58] show these structures admirably.

Above the upper jaw the lower part of the head consists of a mass of fibrous and elastic tissue permeated with oil, derived by modification of a muscular mass (the nasal plug muscle) which was known to the old whalers as the 'junk'. Above this lies the 'case', the muscular and other tissues surrounding the spermaceti organ, a looser mass of fibrous and elastic tissue filled with spermaceti which is a clear liquid that sets to a solid white wax on cooling. The spermaceti organ originates on the right side of the nasal septum, and attains so great a size that the right nostril does not reach the surface or function for breathing, but is modified as a series of air sacs. The left nostril passes obliquely left of the case to open on the left side at the tip of the head by a curved blow-hole shaped like the hole in the belly of a fiddle. Complex air sacs branch off from both nostrils; their possible functions are discussed in later chapters.

The sexes of the Sperm whale differ widely in size; the females barely reach half the maximum length of the males – few other whales show so great a disparity between the sexes. They are much the same in colour; the black of the upper parts shades off into grey below, where there are often patches of white, though there is much individual variation. In

33

males the bulbous front of the head is generally flecked with grey patches which tend to become confluent with increasing age.

The Pygmy Sperm whale, *Kogia breviceps*, well deserves its name, for the males only attain a length of about 13 feet and the females no more than 9 or 10. Its appearance is little like a Sperm whale's, for its general shape is like that of a porpoise, and it has a dorsal fin and pointed not rounded paddles. Its internal anatomy, however, shows its relationship to the larger species. Only the lower jaw carries teeth, which fit into hollows of the upper jaw when the mouth is shut; furthermore, the lower jaw, like that of the Sperm whale, is shorter than the upper so that it does not reach the end of the snout. In nearly all the other Cetacea, if there is a difference in the length of the jaws, the lower jaw undershoots the upper. In the skull the maxillae and supraoccipital bones form a crest enclosing a bowl-shaped hollow, but the forward extension of the maxillae and premaxillae is short, so that there is not a long shelf-like rostrum as in the Sperm whale; the individual bones forming the bowl are strongly asymmetrical. The right nostril forms a spermaceti organ, and the left one alone functions for breathing, opening at a crescentic blow-hole on the left side of the head towards the snout. The shortness of the skull is correlated with the absence of a huge head made up of case and junk, which are much smaller than in the Sperm whale, and with the general porpoise-like appearance of the animal. The colour is black above shading to greyish below. Several species of *Kogia* have been named, but some probably merely represent differences in age and sex.

The family Monodontidae, the Delphinapteridae of some systematists, contains only two species, the White whale and the Narwhal, both found only in the Arctic and northern oceans. They are small whales, neither species reaching a length of much over 15 feet from snout to tail, and neither has a dorsal fin. The front of the head is rounded, and in the White whale the tip of the upper jaw projects slightly below the melon though not enough to form a beak. The paddles are rather rounded and broad.

The White whale starts life coloured grey, but as it grows the colour fades through yellowish to pure white when it is fully adult at the age of about five years. Both upper and lower jaws are armed with eight to ten teeth. This whale is also known as the Beluga, a Russian name not to be confused with the similar sounding, and in English similarly spelt, Russian name for the great sturgeon from the eggs of which caviar is prepared – in both names the first syllable means 'white'. Belugas inhabit the waters of the shores of the Arctic Ocean and the colder waters of lower latitudes as far as the St Lawrence River in the western Atlantic; they sometimes go up the river as far as Quebec. They are said usually to go about in small schools, but aerial photographs show large schools of several hundreds cruising near the ice edge of Lancaster Sound in Hudson Bay, and elsewhere in the Arctic [190]. In captivity it is a docile and gentle creature. 2056253

The Narwhal is unique in having a unicorn's horn projecting from its snout. The horn is in fact a tooth; both sexes have only two teeth, in the upper jaw, but in the females neither of them erupts from its socket. In the males, however, although that of the right side remains unerupted, that of the left grows out as a spectacular tusk covered with twisted grooves forming a left-handed spiral pattern. Although the horn is a tooth it does not project from the mouth but comes out through a hole in the upper lip just above it. Very rarely the right tooth also grows out so that the animal has two tusks, and it is peculiar that when this happens the right tusk is also a left-hand spiral, not as might be expected symmetrical and right-handed. Even more rarely a female grows a tusk, probably because of some abnormality in her hormones. The asymmetry is so surprising that when Sir Richard Owen, the Victorian anatomist, was shown a two-tusked Narwhal skull in a Netherlands museum he declared that it was a fake. The use, if any, that the male Narwhal makes of its tusk is unknown – zoologists say it is a secondary sexual character of the male, a short way of saying only males have a tusk. It has, however, recently been stated [3] that divers working on an oil rig off Baffin Island saw Narwhals use their tusks to 'rip trenches for food in the ocean floor'. But we are not told what food the animals were seeking, nor how competent the oil divers were as observers and naturalists. If the report is correct it leaves unanswered the question of what food the females eat and how they find it.

The colour of the Narwhal, like that of the Beluga, varies with age. At birth it is grey all over but as the animal grows up the belly becomes white and the back black, covered with black or white specklings and small patches. The Narwhal is confined to the Arctic, and does not frequent shallow water and river estuaries as does the Beluga.

The large family Delphinidae is divided into at least fifteen genera by the most parsimonious systematists. The number of species among these genera is great but not exactly known because some are rare so that odd specimens that have been found at different places and different times, though given specific names, may not really be distinct. On the other hand some systematists are coming to regard separate populations of some widespread forms that were thought to be conspecific as separate species rather than as races or subspecies.

The family contains the dolphins and porpoises apart from those of the family Platanistidae; they are mostly small to medium-sized cetaceans, ranging in length from about 5 to 15 or even 20 feet – the giant among them, the Killer whale, reaches 30 feet in length. The names 'porpoise' and 'dolphin' are vernacular words and consequently not generally precise; English cetologists tend to call the beaked species dolphins and those without a beak porpoises, but they are not consistent. American usage differs; for example the animal called Bottle-nosed dolphin in English is the Bottle-nosed porpoise in America. Confusion can also arise because a large brightly coloured oceanic fish is also, correctly, named the dolphin; it is famous for the range of colour changes it shows in its death pangs. The Common dolphin and the Common porpoise of European and other seas can be assumed to be the animals to which the names originally belonged.

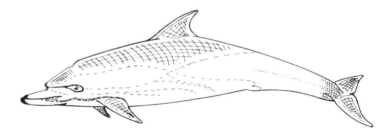

Figure 5. Common dolphin.

Most of the delphinids, but not all of them, have functional teeth in both jaws, and similarly most of them have a back fin. All of them have a single crescentic blow-hole with the concave side facing forwards on the top of the head, but in none of them are the bones of the skull raised into crests similar to those of the Sperm and Beaked whales. The colour is basically dark above and light below, but there is great interspecific variation, often with striking patterns of black and white; some species include shades of brown, yellow, pink, blue or grey in their coloration. In all the delphinids the melon is present, and in some species it forms a conspicuous bulging forehead.

36

The colour patterns of the delphinids have been classified by the Canadian cetologist Edward Mitchell [125], who elaborated the similar studies of the Russian A. V. Yablokov [214]. He regards the simple dark-above-light-below *saddled* pattern as the basic from which the *spotted*, the longitudinally *striped*, and the *crisscross* ones are derived, as also are the piebald patterns of striking black and white with the addition of distinct blazes of white on black in various positions. Yablokov made three basic types, uniform, patched, and coun-tershaded, and suggested that the more conspicuous patched patterns are useful in recognition between the animals, the uniform ones are related to feeding at depths where light is dim, and the countershaded ones to feeding near the surface. None of the patterns serve as pro-tection from predators. Mitchell agrees with these ideas but suggests further that though the saddled and spotted patterns are probably concealing, the blotched patterns are probably disruptive as well as serving for recognition, and that the crisscross pattern, which is the most highly developed, serves as disguise through disruptive pig-mentation and camouflage by countershading and shadow mimicry. He also emphasizes the importance of the colour patterns in studying the phylogeny of the delphinids.

The genus *Steno* probably contains only one species, *S. bredanensis*, the Rough-toothed dolphin; though specific names have been given to several specimens, they all appear to belong to the one species. The tips of the teeth are roughened with furrows; and the beak, though long, is not sharply marked off from the head, which has a gently convex profile from snout to back. The upper parts are greyish-black with irregular light markings, and the under parts pinkish-white with grey splashes. The Rough-toothed dolphin is known from the Indian Ocean and the tropical and subtropical Atlantic.

Sotalia is a genus containing several species of small to medium-sized dolphins which mostly live in tropical coastal waters, estuaries and rivers. They have rather long beaks well marked off from the head, and numerous teeth; the colour ranges from dark grey to greyish-yellow. Three species have been named from the upper Amazon, but they are probably merely colour variations of a single one; their native names are 'buffeo' and 'tucuxí', pronounced too-coo-shee. Two other species, if indeed they are distinct, inhabit the coastal waters of tropical South America, *S. guianensis* and *S. brasiliensis*.

The several imperfectly known species of *Sousa* were formerly classified in the genus *Sotalia* but are now separated on the character of the ear bones. They are known from the tropical Atlantic coast of Africa, and from the Indian Ocean and Chinese waters. The beak is well defined and carries numerous teeth, but the most striking feature of these dolphins is great development on the back of the hump bearing the dorsal fin, and the depth of the dorsal and ventral keels on the tail-stock, at least in the adult males. Hence the names of the two known species – the Atlantic Hump-backed dolphin, *S. teuszii*, and the Indo-Pacific Hump-backed dolphin, *S. chinensis*.

The genus *Stenella* – the name supersedes *Prodelphinus* formerly used – contains a large number of names applied to an uncertain number of species, certainly less. The genus seems to be a systematist's nightmare, for many species have been described from the examination of single skulls with no attached information about the appearance and coloration of the whole animal. Furthermore, collections in museums contain many skulls of animals in this genus which do not fit the description of any species, but are intermediate and indefinite [64]. The species known in the flesh vary in coloration from uniform grey to the black above and white below, spotted and splashed with darker and lighter, of *S. plagiodon* of the western Atlantic. *S. longirostris* and *S. coeruleoalba* have more distinctive colour patterns: the back in the first is black, in the second dark blue; in both, the under parts and belly are white. A narrow ribbon of black or blue runs from a ring round the eye along the side and broadens at the vent, and another band runs from the ring to the base of the flipper. In *S. coeruleoalba* a dark stripe runs forward from the back fin. These patterns give the animals a most handsome appearance, and the ring round the eye gives a facial expression of intelligence attractive to human eyes – and completely illusory. All the species of *Stenella* are of moderate size, ranging from about 6 to 9 feet in length. The beak is long and narrow and the teeth are pointed and small, so that in some species they number over fifty pairs in each jaw. The animals live in tropical and subtropical waters; about half a dozen species are known from the Atlantic, Pacific, and Mediterranean.

In the genus *Delphinus*, too, a large number of species has been named in error, for most of them belong to a single widespread species, the Common dolphin *D. delphis*. It reaches a length of not more than a few inches over 8 feet; its beak is well marked and narrow, with up to fifty pairs of small teeth in each jaw. Ill-defined stripes of yellow, grey and white merge together in an irregular pattern on the sides between the black or dark brown of the back and the white of the underside. A black patch surrounds the eye from which a dark line runs towards the beak. The Common dolphin is found, often in large schools, in all except the colder seas of the world.

Risso's dolphin, *Grampus griseus*, the only species of its genus, is a much larger animal, reaching a length of 13 feet. The jaws are not prolonged into a beak, or perhaps it would be more correct to say that the forehead extends nearly to the end of the snout. The upper jaw is toothless, and the under jaw carries only three to seven comparatively large teeth near its tip. The flippers are long and narrow, and the dorsal fin is rather high. The colour is unusual: most of the body is grey merging irregularly into darker and lighter areas, the darkest on the back and lightest on the head and belly. A large number of lines, many of them in parallel series of two or three, is imposed on the background, some of them·nearly straight, others curved or even 'scribbled'. A smaller number of light spots is scattered among the streaks. Risso's dolphin is widely distributed, being known from the Atlantic and Mediterranean and from the south Pacific. It is said to go in small schools of less than a dozen, but in 1948 I sailed through a school in Bantry Bay which I estimated to contain at least twenty animals.

The genus *Tursiops* contains the species best known to the general public because it is the one usually shown in the water circuses of the large marine aquariums. *T. truncatus*, the Bottle-nosed dolphin, is a large dolphin, reaching a length of about 12 feet; it has a short stout beak with twenty or so fairly large teeth on each side of both jaws. The back is black or dark grey and the belly white, as is the lower jaw. This dolphin is common in the Atlantic – it was formerly fished commercially at Cape Hatteras – and in the south Pacific; other species of the genus are known from the Red Sea, the Indian Ocean and Australia.

There are about six species in the genus *Lagenorhynchus*, some of them imperfectly known. They are medium-sized dolphins ranging in length from about 6 to 10 feet. The beak is short, with about twenty-four to thirty-four teeth according to species. The tail bears a longitudinal ridge on the upper and lower surfaces between the flukes and the back fin, and the vent. All the species have a distinctive colour pattern, which is a striking contrast of black and white patches and streaks in most of them. The English names of some species refer to the colours, such as the White-beaked and White-sided dolphins, as does the Latin name of *L. cruciger*. The genus is widely distributed, different species being found in all the oceans from the Arctic to the Antarctic. Some species are gregarious and sometimes go in enormous schools. *L. hosei* is placed by some systematists in a separate genus *Lagenodelphis* more closely related to *Stenella* and *Delphis*.

Most of the dolphins in the genus *Cephalorhynchus* are also distinguished by striking patterns of black and white. They are small dolphins, none of the four named species exceeding 6 feet in length. The beak is very short and not sharply marked off from the rest of the head, and has some twenty-five to thirty small teeth in each jaw. The genus is confined to the southern hemisphere, the different species being known from the southern oceans all round the world. Commerson's dolphin, otherwise known as the Piebald porpoise, *C. commersoni*, carries the most striking pattern of sharply defined black and white areas.

The Killer whale, *Orcinus orca*, the largest of the dolphins, is likewise strikingly patterned. The upper parts are black, the lower white; there is a white patch on the side of the head above the eye, and another on the side of the back below and behind the back fin. Behind the level of the back fin the white of the belly extends up on the flank as an oval patch of white sharply defined in the surrounding black. The Killer does not have a beak, the profile of the head thus resembling that of a porpoise. Each jaw bears ten to twelve large conical teeth on each side. This species differs from all the other odontocetes, except the Sperm whale, in the great discrepancy in size between the sexes; females reach a length of about 15 feet but the males grow to twice that length, so that they are indeed giant dolphins. In the females and young males the back fin has a concave rear edge – the fin is falcate – but in adult males it becomes triangular and greatly exaggerated in height, up to 6 feet or more. The paddle or flipper is rounded in outline, and in the adult males it grows to a huge size so that its length may reach nearly a quarter of the entire body length. This enormous and apparently disproportionate overgrowth has been said to be correlated with high swimming speeds and predatory activity. As the Killer goes about in schools of up to thirty or forty animals of mixed sexes, and the females, which do not enjoy this bizarre distortion, are able to keep up with the males, this explanation is not convincing. It may be that the overgrowth is the result of allometric growth like that which produced the enormous and probably useless antlers of the extinct pleistocene giant deer,

Megaceros. There is only one species of Killer: it inhabits all the oceans of the world.

Similarly the genus *Pseudorca* contains a single species *P. crassidens*, the False Killer whale. It is much smaller than the true Killer, approaching but not reaching a maximum length of 20 feet. As in the Killer, the snout is not prolonged into a beak but has a rounded profile with slightly underhung lower jaw. The stout conical teeth number eight to eleven on each side of both jaws. The body shape is slimmer than that of the Killer; the dorsal fin and flippers are smaller in proportion, the latter being elongated but not rounded. The colour is uniform black all over. The False Killer is an oceanic species but occasionally it comes to inshore waters, an environment beyond its usual experience, where it comes to grief by getting stranded, often in large schools numbering up to several hundred. From these sporadic mass strandings it is evident that the False Killer inhabits all the oceans worldwide.

The Irrawaddy dolphin, *Orcaella brevirostris*, the only species in its genus, is also beakless, and in shape of body resembles a miniature False Killer 7 to 8 feet long. The flipper, however, is proportionately broader and rounder, and the back fin is small and rounded; the colour is slaty-grey, a little lighter on the under parts. The Irrawaddy dolphin is a tropical species that lives in both salt and fresh water; it is found in the coastal seas of Indonesia, north Australia, and Indo-China, and in the larger rivers of those regions, ascending as much as 900 miles from the sea in the Irrawaddy.

The genus *Feresa* contains a single species, *F. attenuata*, which much resembles a small False Killer, so that it has been named the Pygmy Killer whale. Until recently it was known only from a few skulls in museums, but schools of up to fifty have been seen off Hawaii and a few specimens have been captured for examination there and on the coast of Japan.

Similarly, *Peponocephala electra*, a formerly rare species, the sole member of its genus, has in recent years become better known as a result of the stranding of large schools. Both species have been found in tropical and warm temperate seas.

The generic name of the Pilot whale, *Globicephala*, describes its most striking character, the large rounded melon making a lofty 'forehead' above the mouth. The edge of the upper jaw projects as a narrow shelf below the melon, but not enough to form more than a rudimentary indication of a beak. About ten moderately stout teeth grow in the front part of the jaws. The long, low, backwardly hooked back fin is also characteristic, as also is the long narrow flipper. The Pilot whale is a large dolphin, the males reaching 18 feet in length, the females about 15; the colour is black all over with a heart-shaped white patch on the throat in one species – the old whalers who hunted them at sea called them 'Blackfish'. Pilot whales are found in all seas, those of the northern and southern regions being the species *G. melaena*, those of the warmer and tropical seas *G. macrorhynchus*, which has a proportionately shorter flipper. Pilot whales are highly gregarious, often swimming in schools of several hundred. This habit is exploited by men for making large catches by driving a school ashore so that the stranded whales can be slaughtered. This method is still used in Faroe and Newfoundland, and was formerly used in several other places such as Orkney and Norway. The Scotch vernacular name, Caa'ing whale, is derived from the hunting method of driving whales ashore.

The two species of Right whale dolphins, genus *Lissodelphis*, are so called because, like the Right whale, they do not have a back fin. There is a short beak with over forty pairs of small pointed teeth in each jaw. The coloration is distinctive: a pattern of black and white areas with sharply defined edges: in the southern species, *L. peroni*, which reaches about 6 feet in length, the underside is white and a cloak of black covers the back from a little above the beak to the tail, running high along the side but reaching towards the root of the flipper. In the northern species, *L. borealis*, which reaches a length of about 8 feet, the black is more extensive; the largest white area is a patch on the belly between the flippers, from which a white ribbon runs to the tail, and a thin line forwards to a small patch under the end of the beak – the outer parts of the flukes also are white on the underside. These beautiful dolphins are found in cold waters, the southern species throughout the southern ocean, but the northern species only in the Pacific.

The several species of the genus *Phocoena* include the Common porpoise, known as the Harbor porpoise in the United States of America. The porpoises are small cetaceans, not more than about 6 feet in length and rather stoutly built; the profile of the snout is rounded without any trace of a beak. The teeth differ from those of most other genera in being not conical but spade-shaped, and range in number from sixteen to about two dozen, according to species. The Common porpoise, *P. phocoena*, black above and white below, is common in the

Atlantic and Pacific Oceans from the Arctic to the northern tropic. The other species have more limited distributions; one, *P. sinus*, is confined to the Gulf of California; Burmeister's porpoise, *P. spinipinnis*, entirely black, and with a row of tubercles on the leading edge of the dorsal fin, is found round the coasts of South America from Uruguay to Peru; and the Spectacled porpoise, *P. dioptrica*, with the black of the back meeting the white of the flanks and belly in a hard line resembling that of the Right whale dolphins, inhabits the south Atlantic from Uruguay to the sub-Antarctic. Dall's porpoise is placed in a separate genus, *Phocoenoides*, as it differs in having very small teeth and a larger number of vertebrae. *P. dalli*, about 5 feet long, black above with white flanks and belly, is abundant in the north Pacific. Finally, to conclude this summary of the delphinid genera, the small Finless porpoise, *Neophocaena phocaenoides*, delicate grey but lighter below and lacking a dorsal fin, is common in the Indian Ocean, being found from the Cape of Good Hope to Japan, and often ascends estuaries and rivers. It has a row of small tubercles on the back where the dorsal fin lies in other species.

The whalebone whales, the Mysticeti, few in species but enormous in individual bulk, remain to be considered. These whales differ from the others in many ways, the most obvious being the baleen apparatus in the mouth and the correlated way of feeding. Baleen is a fibrous, horny flexible material that grows as a series of flat plates, each about a quarter of an inch thick, from the sides of the upper jaw in the position of the upper teeth in other animals. The plates are scalene triangles with two long and one short side; they are set across the long axis of the mouth with the short side forming the base. The outer edge is smooth but the inner edge is frayed out into a fringe of fibres. In some species there may be over four hundred plates spaced about a quarter of an inch apart in a row on each side of the mouth. The frayed inner edges cover the spaces between the plates as a fibrous mat or filter bed, which is used for straining from the sea water the planktonic creatures, from small copepod Crustacea to fishes the size of herrings, that form the food. Water loaded with plankton enters the mouth and is expelled between the plates, leaving the food stranded on the mat, thereafter to be swallowed. In the rorquals and the Humpback the fibres form a series of hairs at the front, bridging the gap between the sides.

At first sight it is difficult to imagine how this beautifully adapted apparatus can have evolved, at least by the generally accepted theory that evolution proceeds by the 'natural selection' of the effects of random changes in the genes. The baleen is an entirely new structure and has nothing to do with teeth – rudimentary teeth or tooth-buds that come to nothing are present in the embryo. The gradual suppression of teeth and the acquisition of baleen seems improbable, for the baleen could apparently function only when it is fully developed. Once a useful baleen apparatus has been attained evolutionary changes could understandably improve its efficiency; but it seems impossible that a fortuitous change in the genetic code could produce a workable structure, with all the attendant changes in physiology and behaviour, at a stroke. Would the first whalebone whale thus evolved know what to do with its baleen? Further consideration, however, shows that baleen was indeed evolved gradually through the modification of already existing structures.

In the first place the baleen does not 'grow' from the gums but from the greater part of the palate in the roof of the mouth; the visible palate in baleen whales is a narrow strip in the centre between the two sides of baleen. If we examine the palate of other mammals is it possible to find any structures that might be regarded as precursors of the baleen in the mysticetes? The affirmative answer is provided by the very common occurrence of transverse ridges of the tough lining of the roof of the mouth in many land mammals. The ridges facilitate the manipulation of the food during mastication and swallowing; they can be easily seen in the mouth of domestic animals, but not in ourselves for the higher primates do not have them. If we look at the hard palate of a cat we see it is crossed by seven or eight curved transverse ridges – in the dog there are nine or ten with an ill-defined longitudinal ridge in the centre dividing the two sides. In the horse there are eighteen to twenty on each side divided by a more prominent central ridge. These curved cross-ridges are held to be the equivalent of the baleen plates of the mysticetes, and the central ridge the equivalent of the exposed part of the palate between the baleen sides.

This supposition is strengthened by the condition found in some abnormal Sei whales examined by the American cetologist Dale Rice at whaling stations in California in 1959 [168]. He found two, out of thirty-nine, Sei whales in which the baleen was so rudimentary as to be almost lacking, and a third in which it was rudimentary on one side. The baleen plates were only 2 to 7 centimetres long and were without a hairy fringe so that they were completely non-functional for straining plankton from the water. The plates were unusually soft and thick and were easily pulled away from their attachment. All the whales were fully adult, aged from ten to fourteen years, and one female was lactating; all were in good condition with blubber of the normal thickness for the time of year. The stomach of one was empty, but that in each of the other two was filled with northern anchovies, so although the animals could not feed by filtering plankton they evidently made a good living by feeding on small fish. Dr Rice was unable to determine whether the abnormal rudimentary condition of the baleen was of genetic origin or due to injury or disease. Whatever the cause, it shows that baleen could have evolved from a proliferation of the ridges of the hard palate present in many mammals, and that in its rudimentary stages it could have helped in the capture of food grabbed by the mouth until with increased development it was used in a different way as a filter.

The elongation of the jaws that accompanies the shifting of the nostrils to the top of the head in all whales also enables the mysticetes to carry the long series of baleen plates – up to more than four hundred and fifty on each side in the Fin whale. The maxillae and premaxillae form a long rostrum or snout in front of the cranium or brain box, which latter is roofed mainly by the enlarged supraoccipital bone of the back of the skull, so that the parietals and frontals are in effect squeezed forwards and outwards to take a minor part in roofing the brain case. Small nasal bones are present because the two nostrils remain separate, so that the blow-hole is a double hole forming two parallel slits close together when shut. In the skull, unlike that of the toothed whales, the component bones are not asymmetrical; as Beddard [14] said portentously in 1900 'the skull is nearly symmetrical; in fact, it is not perceptibly asymmetrical'.

The living species of the Mysticeti are classified into three families. The Balaenidae contains the Right whales, in which the baleen apparatus reaches its greatest development; the Balaenopteridae contains several species of rorquals and the Humpback whale, in which the baleen is comparatively short and the throat is covered with a system of grooves that have been likened to tramlines; and the Eschrichtiidae contains only the Gray whale, with short baleen and no tramlines.

In the Right whales the head is large, in some specimens up to one-third the total length of the body, and the rostrum is long, narrow and arched upwards. The lower jaw-bones are not similarly arched, so that the space between upper and lower jaws is closed by the huge lower lips that rise from the jaw-bones and enclose the long narrow baleen plates hanging from the edges of the rostrum. The large species of Right whale do not have a dorsal fin; their colour is mainly black with variable amounts of white in patches on the underside; and they are up to 60 feet

Figure 6. Black Right whale.

in length. The Greenland Right whale or Bowhead is confined to the Arctic, but the Black Right whale, of which numerous geographical races or separate populations have been named, is cosmopolitan in the oceans between the Arctic and the Antarctic circles. A small species, the Pygmy Right whale, not exceeding 30 feet in length, differs in having a back fin and relatively shorter baleen, but has the seven neck vertebrae fused into a single mass, thus resembling the other Right whales to which this character is peculiar, for the fusion is never so complete in any of the other whales. It is a rarely seen species found in southern temperate regions.

The balaenopterid whales are slimmer in build and less bizarre in the size of the head and elaboration of the baleen than the Right whales. They all have throat grooves and a dorsal fin; and the rostrum is much broader and flatter, with shorter and wider baleen, than in the Right whales. The five species of the genus *Balaenoptera* are known collectively as rorquals or finner whales, though the name Fin whale applies scientifically to one. They range in length from a maximum of about 100 feet in the Blue whale to about 30 feet in the Piked whale, now generally known by its Norwegian name of Minke whale. The Blue, Fin, Sei, Bryde's and Minke whales bear a strong family resemblance to

Figure 7. Blue whale.

46

each other and, apart from size, differ in details of colour, size and shape of dorsal fin, and number of throat grooves and baleen plates. The Humpback whale differs enough to be placed in a separate genus, *Megaptera*. It reaches a length of only about 50 feet, but is much more stoutly built than the rorquals so that its oil-yield is little less than that of a Fin whale 70 feet long. Its colour is black above and white below but there is great individual variation in the relative amounts of the two. The throat grooves are fewer but much coarser than in the rorquals and the flipper is exceedingly long and narrow. The head, lower jaw and edge of the flipper are covered with irregular knobs, and the trailing edge of the tail flukes is indented and serrated. The small dorsal fin is set rather far back towards the tail. The rorquals and the Humpback are found in all the oceans of the world.

The Gray whale, *Eschrichtius robustus*, is placed in a separate family because it is in some ways intermediate in structure between the Right whales and the rorquals. The rostrum is narrower than that of the rorquals and gently arched, though less than in the Right whales. There are two, very rarely four, short throat grooves, and there is no dorsal fin, but the hind end of the back bears up to ten low humps towards the tail. As its name shows, its colour is grey; although it is a small whale not exceeding about 45 feet in length, it is very active, and so aggressive when attacked that the old Californian and Japanese whalers knew it as the 'Devil Fish'. It is found only in the north Pacific, on the east side from Lower California to the Arctic, and on the west side from Japan and Korea to Kamptchatka, though the Korean herd is now probably extinct.

We have now briefly reviewed the diversity of whales as a basis for examining in more detail some of the fascinating aspects of their biology, adapted to a way of life in a medium impossible to the majority of mammals. The solution of the problems facing an air-breathing animal in an aquatic environment has enabled the whales to exploit on a large scale the vast food resources of the sea.

Chapter 3

Food and feeding

The basic activities of all animals are, first, to find their food, and, secondly, to reproduce their kind. In this chapter we examine the first subject, and its ecological background.

All cetaceans are carnivorous in that they eat other animals, although few in fact subsist on flesh as do most of the land carnivores. The mysticetes and the odontocetes, so different in their physical structure, differ equally in their diets, the former being grazers in bulk on swarms of comparatively small creatures, and the latter taking food items of larger size one at a time. The baleen apparatus of the mysticetes is a highly specialized arrangement used for the capture of food by filtering it from sea water, whereas the dentition of the odontocetes is simplified by being in general homodont and, in many species, reduced. In Chapter 2 we have seen how greatly the dentition differs in the various genera, and that the teeth are very numerous in some but absent in others, at least as functional structures.

The specialized structures in the mouth used for catching the food are not the only unusual features in the food canal of the cetaceans: the form of the stomach differs from that of other mammals. This organ, unlike the comparatively simple sac found in many orders of the Mammalia, and particularly in the true carnivores, is divided into at least four compartments – more in some species. The comparative anatomy of the stomach was investigated in detail by the cetologists of the nineteenth century and culminated in an authoritative memoir by Jungklaus published in 1898 [92]. In all cetaceans except the ziphiids the first chamber of the stomach is a dilatable sac-shaped extension of the oesophagus; its lining, like that of the oesophagus, contains no digestive glands. The second chamber lies alongside the first because the opening into it adjoins that from the oesophagus; its lining mucosa is much thicker and is deeply folded so that its internal surface is thereby increased. This mucosa is richly supplied with digestive glands [87]. A small third chamber, in some species no more than a dilation in the wall of the second, connects the second with the much larger fourth chamber. The last is the pyloric part of the stomach; it is more intestiniform and is folded on itself. In many species it is subdivided by constrictions and

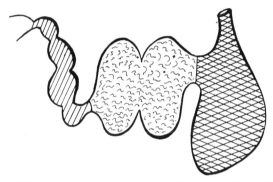

Figure 8. The multiple-chambered stomach of a Fin whale. The gullet leads to the first chamber on the right.

semilunar valves, especially in the ziphiid whales in which the total number of chambers may reach thirteen or fourteen, although the first, oesophageal chamber is rudimentary. In many species there is another sac-like dilation distal to the pyloric constriction of the terminal chamber of the stomach. This is, however, not another stomach compartment but the beginning of the duodenum, as is shown by the common pancreatic and bile duct which discharges into it. The complex stomach and the poor development of salivary glands is correlated with the fact that cetaceans swallow their food without masticating it.

The intestine of the whales is comparatively long and in most species of toothed whales shows little differentiation into small and large intestine, though in the mysticetes the cæcum and colon are distinct. The length of the intestine in the mysticetes is five or six times the length of the body, but in some of the toothed whales such as the Sperm whale and Bottle-nosed dolphin it reaches fifteen or sixteen times the body length [24].

The first chamber of the stomach is in effect a dilation of the oesophagus, and may thus be regarded as analagous with the crop of birds and to have a similar function. It has, nevertheless, been claimed that some digestion takes place in it by digestive juices leaked back from the second compartment – in the porpoise [62], the Killer [53], and the whalebone whales [215]. It does not appear to have been proved that this digestion takes place during life, and the anatomical arrangement of the parts and the valves at the junctions of the chambers suggest that it may occur only after death. The specimens of Cetacea dissected by naturalists are seldom freshly killed; they have usually been found stranded or floating at sea some time after death – the Killer in the stomach of which Professor Eschricht found the remains of numerous porpoises and seals in 1861 had been found at sea, and was in such a state of decomposition that he feared the stench would overcome him and his assistants before they could prepare the skeleton for transport to the Copenhagen museum. In as short a time as three to four hours the krill eaten by a baleen whale that has filled its stomach just before being killed is reduced, as Zenkovich says [215], to the consistency of 'a thin gruel'.

These observations show that digestion goes on after the death of a whale, probably by regurgitation of gastric juice into the first chamber of the stomach, but they neither prove, nor exclude the possibility, that digestion takes place there during life. Normal decomposition, in addition to digestion, may help the dissolution of food remaining in the stomach after death. The insulating effect of the blubber in retaining heat in the carcase leads to a quick rise in temperature with rapid production of the gases of decomposition, so rapid that the extensible throat of baleen whales blows up like a balloon within twelve hours of death. Decomposition in the odontocetes, which have no similar weak part of the body wall, bloats the whole carcase and produces such a high internal pressure that whalers always used to open the abdomen of a Sperm whale that could not be cut up at once on being delivered to a whaling station, to relieve the pressure and avert a dangerous explosion.

We have seen how the baleen of the mysticetes acts as a strainer to filter the food from the sea water, but before discussing the food and feeding habits of these whales let us look briefly at the structure of that remarkable substance.

The frayed inner edges of baleen plates resemble coarse hair or bristle not only in superficial appearance but in origin. Both, like the enamel of teeth, are derived from the embryonic ectoderm, the cell-layer that gives rise to the skin and its derivatives. As long ago as 1787 John

Hunter [90] showed that a baleen plate consists of rows of long hairs stuck together and covered by a layer of horn except at their outer ends, where they are separated to form the fringe. A mass of fleshy connective tissue attached to each side of the upper jaw supports the series of plates; its buccal surface is raised into ridges or basal plates, each of which gives rise to a baleen plate. The basal plates carry a number of long slender conical papillae, and are themselves formed by the fusion of rows of papillae. The layer of epidermis covering the papillae, the basal plates, and the spaces between the plates, produces the horny substance of which baleen is made. The epidermis of each papilla produces a long tubular bristle or horn tube, and that of the spaces between the papillae produces layers of compacting horn which stick the tubes together. The outer ends of the horn tubes project beyond the compacting and covering layers of horn as the bristly fringe. The spaces between the plates also produce a horny substance which covers the compacted horn tubes, and in addition produces successive layers of softer horn between adjacent baleen plates so that their bases appear to be embedded in it and anchored to the jaw by it. When a baleen whale is cut up by whalers the entire 'side' of baleen on each side of the upper jaw is cut off by dividing the underlying connective tissue so that the plates remain attached to each other by the white intermediate substance of soft horn. In the days when the very long narrow baleen plates of the Right whale were the most valuable product of the animal the plates were separated and the adhering intermediate substance of 'gum' was scraped away.

The size and shape of the baleen plates in the Right whales differs widely from those of the rorquals, being extremely narrow, long and flexible. The plates of all baleen whales are longest at the middle of the series, diminishing in size towards each end. The plates of the Right whales are only up to about 1 foot wide at their bases, usually less, but grow to a great length, tapering to a point. They are longest in B. mysticetus, the Greenland Right whale, the extreme recorded by Scoresby [184] being 15 feet long, though plates 12 to 13 feet long were regarded as uncommon. Blades over 6 feet long were called 'size bone', and were sold separately from the 'under-size', which fetched only half the price.

In the rorquals the baleen is shorter and wider. The width of the plates in the largest of the rorquals, the Blue whale, B. musculus, reaches nearly $2\frac{1}{2}$ feet and ranges upwards from three-quarters of the length rarely to equal it; the greatest length is about 3 feet. It thus approximates in shape to a right-angled triangle, in contrast to the long, strip-like narrow triangle of the baleen plates in the Right whales. In the other rorquals, none of which reaches the size of the Blue whale, the baleen plates are correspondingly smaller, the texture of the hairy fringe differs in the different species, and in all the plates are stiffer and much less flexible than those of the Right whales.

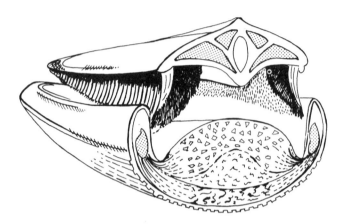

Figure 9. Diagrammatic cross-section of the head of a baleen whale. The baleen plates hang from the edges of the upper jaws, and the tongue lies between the lower jaws. The tip of the snout is to the left.

The colour of the baleen in Right whales is black with occasional white streaks running lengthways; in the rorquals it varies according to species from black through dark grey to almost white, and in some both dark and light coloured plates occur in one 'side' of baleen, notably in the Fin whale, *B. physalus*. In this species the plates of the fore third of the side on the right are white whereas those of the rest and of the left side are dull grey with streaks of white, but the hairy fringe of all the plates is yellowish-white. All the plates and the hairy fringes of the Blue whale are black; those of the Sei whale are black but the fine silky hairy fringe is white; in Bryde's whale they are mostly black but the fore end of each side is white and the hairy fringe very coarse; and in the Minke whale all the plates, with the exception of some at the back, and the fringes are yellowish-white.

The greater width of the baleen plates in the rorquals corresponds with the greater width of the fore part of the skull, the rostrum, which tapers in a broad triangle from the region of the jaw joint to the tip. In the Right whales the rostrum rises from the jaw joint as a very narrow high-arched strip bowed downwards to the tip; the sides of baleen grow along each side leaving a very narrow palate between them. In the rorquals also the palate is a narrow strip between the baleen sides, not because of the narrowness of the rostrum, as in the Right whales, but because of the greater width of the plates; in Right whales the combined width of the two sides is not more than 2 feet whereas in the rorquals it may be up to 5 feet.

In the rorquals the comparatively short baleen is easily contained in the mouth when it is shut, but in Right whales the blades are about twice as long as the distance between the upper and lower jaws when the mouth is shut. It was not until 1878 that Captain David Gray of Peterhead, one of the most famous Arctic whalers of the latter part of the nineteenth century, at the request of W. H. Flower [61], explained how the baleen plates are folded back towards the throat, the front ones passing below the hinder, so that they lie on each side of the tongue when the mouth is shut. As Flower remarks, this is possible through the elasticity and flexibility of the blades which is not, 'primarily at least, for the benefit of the corset and umbrella makers'. When the mouth opens the plates spring forward and become straight, and the ends are retained within the mouth by the lower lips, which rise several feet above the jaw-bones. The open mouth is like 'an enormous spoon' roofed by the baleen sides rising to the narrow ridge of the rostrum, because the hinges of the jaw-bones are spaced far apart on each side of the skull and the jaws themselves bow out farther before curving inwards to meet at the tip. The Right whales when feeding thus scoop up their food, in contrast with the rorquals which cannot use their straining apparatus with the mouth more than slightly open so that the tips of the baleen plates remain within the lower lips, and gulp their food by depressing the throat to increase the capacity of the mouth.

At first sight it might be thought that the baleen whales feed indiscriminately on mixed plankton, but this is not in fact the case. Many planktonic creatures aggregate in swarms of a kind so that whales can select preferred species. Naturally when feeding on such swarms the whales must take all other organisms that may contaminate the swarms, but generally one species predominates in the bulk food.

Few modern naturalists have had the chance of examining the food of the Greenland Right whale as it is rare and protected from commercial whaling, though taken by the Esquimaux and others who use its products for their own subsistence. The books are generally vague in their statements about its food, merely saying that it consists of 'minute pteropods and crustaceans' – such information seems to be no more than guesswork based on the knowledge that such creatures abound in the Arctic plankton [61]. Scammon [175] refers to the 'animalcules' on which the whales feed and says the whalers call them collectively 'Right whale feed or "brit"'; he later terms them 'insects'. Scoresby [184] however, gives some definite information; he describes and gives illustrations of several kinds of planktonic animals of the Arctic – two species of pteropods, arrow worms, ctenophores, Crustacea including a copepod, a filamentous diatom chain, and others. He says that the food of the whale consists of 'various species of actiniae, cliones, sepiae, medusae, cancri, and helices', but then qualifies this statement by adding 'at least, some of these genera are always to be seen wherever a tribe of whales is found stationary and feeding'. He then abandons guesswork and gives something definite: 'In the dead animals, however, in the few instances in which I have been enabled to open their stomachs, squillae or shrimps were the only substances discovered. In the mouth of a whale just killed, I once found a quantity of the same insect.' The illustration of the animal he terms a 'squilla' shows an amphipod crustacean, which in his text he calls *Gammarus arcticus* (Leach) or *mountebank shrimp* from its action of turning over in the water – the species is now known as *G. wilkitzkii*. Scoresby says that it 'inhabits the superficial water, and affords food for whales and birds'. He mentions another of 'this family' under the generic name *Gammarus*, 'remarkable for the largeness of its eyes', a description that leaves no doubt that it was another amphipod of the genus *Parathemisto* or of one of the closely related genera of the family Hyperiidae. Scoresby says that this species also is eaten by the Greenland whale for it 'was found in large quantities in the stomach and mouth of some mysticete'. Hence, although he captions his plate illustrating the planktonic creatures listed above 'Figures of Medusae and other animals, constituting the principal food of the whale' the only one of them he had ever seen in the mouth or stomach of a whale was the amphipod *G. wilkitzkii*, together with the other species with large eyes which he does not figure. Scoresby emphasizes the enormous quantity of plankton in the Arctic Ocean but, although it is possible or even probable that the Greenland whale feeds on other animals such as the pteropods which swarm in immense numbers, he cannot definitely state that they do. A yearling Greenland whale that strayed as far south as Japan was captured in 1969 and brought into Osaka Bay where it was examined by Nishiwaki and Kasuya [136]. Although the animal was alive when it was beached it lived less than twenty-four hours afterwards and was unfortunately so decomposed when the Japanese scientists arrived to dissect it that the internal organs were nearly reduced to a 'muddy fluid'; the stomach was ruptured and the contents unrecognizable, but the zoologists made sure that it had not contained the bones of any fish or other animal.

The Greenland Right whale is none the less probably an opportunist about its food, as are most animals; it will choose a preferred food if there is a choice but in its absence will take what it can get. The cosmopolitan Black Right whale certainly does; although it is comparatively rare in most of the places where it was formerly common before over-fishing reduced its numbers, we have rather more information about its food and feeding habits, apart from the vague generalities about 'minute pteropods and Crustacea' of so many books.

When Antarctic whaling began with the establishment of a whaling station at Grytviken in South Georgia in 1904 Dr Einar Lönnberg, the Director of the Swedish Natural History Museum in Stockholm, got Captain C. A. Larsen, the famous Norwegian whaler who started the enterprise, to take an assistant at the museum, Erik Sörling, with his expedition to study the fauna of the island and collect specimens. Sörling worked hard for a year and brought back a large collection including the skeletons and other parts of whales, skins and skeletons of seals and birds, and pickled fishes. Lönnberg examined the collection and published an interesting paper about it in 1906 [106]. During his stay in the south Sörling was able to examine seven Black Right whales brought in to the factory. Lönnberg was mainly interested in the anatomy and taxonomy of the species, but adds some notes on its natural history, including the statement by Sörling that 'its food consists of "kril" (Euphausiids)'. This is not surprising, for the other baleen whales when in South Georgian waters feed exclusively on a single species of euphausian, as described below, and the Black Right whales would be expected to eat the same extremely abundant food. Twenty years later the zoologists of the Discovery Expedition [119] were able to examine three Black Right whales brought in to the whaling stations of the island, and to watch the behaviour of a fourth at close quarters. They subsequently worked on a female and calf killed and brought into the station at Saldanha Bay, South Africa. The stomachs of all except one of the adults were empty but the other, killed at South Georgia, contained some krill (*Euphausia superba*); that of the calf contained milk.

Antarctic euphausiid krill is confined to the seas south of the Antarctic convergence, but Black Right whales are not; they consequently feed on other food in other places. In the temperate seas north of the Antarctic convergence, and in the tropics, several species of crustaceans, called 'squat lobsters' in English, are common members of the bottom fauna. They are small creatures ranging from 1 to about 4 inches in length according to species, and look much like tiny lobsters because they carry their claws stretched out in front; the claws are rather slender like those of the Dublin Bay prawn (*Nephrops norvegicus*), now miscalled 'scampi' in English cookery. In life they keep the tail or abdomen tucked forward under the front part of the body or thorax, in a manner similar to that of the common lobster when boiled. In the southern ocean one, or perhaps two species, *Munida subrugosa* and *M. gregaria*, which some zoologists think are the same, occur in large swarms. The females carry their eggs attached to the little 'legs' under the abdomen like the 'coral' of the hen lobster. The larvae that hatch from the eggs, minute and entirely unlike the adult, are planktonic. At successive moults during their growth they increasingly resemble the adults and are recognizably squat lobsters at a length of $\frac{1}{4}$ to $\frac{1}{2}$ an inch, though they do not acquire all the characters of the adult until they are slightly larger, when they leave the plankton and descend to the bottom.

These sub-adult post-larvae were thought by early naturalists to belong to different species or even genera from the adults, and were thus named *Grimothea gregaria* in 1820; now that their true position is known they are sometimes called the grimothea stage of the parent species *M. gregaria*. The grimotheas aggregate in enormous swarms in the plankton, sometimes at the surface, at others down to a depth of at least 50 fathoms. They are bright red in colour so that swarms of them near the surface make wide areas of the sea appear 'as red as blood'. This phenomenon is so striking that it has been recorded by seafarers from Sir Richard Hawkins in 1594 to the present day, particularly off the coasts of Patagonia and the Falkland Islands, and later round New Zealand and its sub-Antarctic islands. Rarely the adults also come to the surface in huge swarms. Swarms of grimotheas in coastal waters often get drifted ashore where their dead and decaying bodies massed at the tide line make an intolerable stench.

Such vast masses of edible protein naturally attract large numbers of carnivorous animals – birds, fishes and others including the baleen whales; the scientists and crew of the RRS *Discovery* also found adult lobster krill most palatable when they were trawling off the Falkland Islands in 1926. During the whaling seasons 1927–8 and 1928–9 the whaling factory ship *Ernesto Tornquits*, Captain Fagerli, took three Black Right whales off the coast of Patagonia; they, like some other baleen whales, had been feeding on lobster krill, for their stomachs were full of grimothea [116]. Up to the last quarter of the nineteenth century there was a flourishing fishery for Black Right whales in the inshore waters of New Zealand, and it can safely be assumed that the swarms of grimothea formed an important part of the food of the species, as it still does of others. In New Zealand the swarms are locally called 'whale feed'.

Small as euphausians and grimotheas may be in comparison with the size of Black Right whales, even smaller creatures of the plankton are regularly eaten. Japanese zoologists obtained special permits from the International Whaling Commission to take Black Right whales in the north Pacific for scientific study in 1956, 1961–3 and 1968 [145]. They took thirteen animals in all, off the coasts of Japan, of Kodiak Island off Alaska and in the Bering and Okhotsk Seas. All these animals had been feeding on copepods, small planktonic Crustacea often called 'water fleas' in English. Professor Omura and his colleagues who carried out the research on the whales conclude that two species of these creatures belonging to the genus *Calanus* 'are the main diet of this species in the North Pacific'.

In the north Atlantic, too, Black Right whales have been found feeding on similar microscopic plankton. During the nineteen years from 1956 onwards the Black Right whales feeding within a few miles of the shore in the waters off Cape Cod in the early months of the year have been regularly watched by the American cetologists Watkins and Schevill [209], who were able to photograph them at close quarters from boats and from low-flying aircraft. The animals feed at the surface on copepod plankton aggregated in floating slicks, and cruise to and fro with the end of the rostrum out of the water. They are thus not gulping mouthfuls and expelling the water through the interstices of the plates with the tongue, but using the mouth and baleen much as a naturalist uses a tow-net for gathering plankton. The whales feed in a similar way when the copepods occur in stratified layers below the surface, and were seen feeding thus at depths down to 30 feet – they probably do so also at greater depths. The whales presumably find the slicks of dense plankton by sight, for they do not skim at random but select the slicks and go from one to another as they eat the plankton in each. On one occasion a whale came within 6 feet of the boat and raised its head out of the water with the mouth wide open high enough to bring its eye above the surface. 'Then it slowly closed its mouth and turned as it backed away and submerged.'

This habit of feeding was recorded by the Hon. Paul Dudley two hundred and fifty years ago [50]. Speaking of the Black Right whale fishery off the coast of New England he says

The Triers, who open them when dead, acquaint me, that they never observed any Grass, Fish, or any other Sort of Food in the right or Whalebone Whale, but only a greyish soft Clay, which the people call *Bole Armoniac*; and yet an experienced Whale-man tells me, that he has seen this Whale in still Weather, skimming on the Surface of the Water, to take a sort of reddish Spawn, or Brett, as some call it, that at some Times will lie upon the Top of the water, for a Mile together.

The 'clay' that looked like Armenian bole was obviously the mass of partly digested copepods in the stomach, and the way the whales obtained it by skimming the plankton at the surface is exactly the same as that seen in the same waters two and a half centuries later.

The rorquals, like the Right whales, are opportunist feeders, but different species have different preferences which are often said to depend on the character of the baleen. South of the Antarctic convergence all the baleen whales eat krill that consists for the greater part of a single species of euphausian, *Euphausia superba*, though some species also take other food. Before discussing this subject, we shall consider what is known of the diet of the different kinds of whale in other seas, for most of them are cosmopolitan in their distribution.

The Blue whale seems to be one of the most restricted species in the choice of its food; in the northern hemisphere it eats krill consisting of euphausians of species different from those of the Antarctic, belonging to the genera *Thysanoessa* and *Meganyctiphanes* – in temperate and subtropical waters it takes little or no food. Off the Patagonian coast, where Humpback, Right and Sei whales were eating lobster krill, Captain Fagerli found that the stomachs of the Blue whales he caught were empty. Similarly in 1926 off Magdalena Bay in Lower California on the Pacific coast of Mexico he found that Sei, Humpback and Gray whales were feeding on swarms of lobster krill but that the Blue whales were not. This tropical lobster krill is a different species from that found in the south, and the adult animals as well as the immature stages commonly swarm at the surface. This rejection of an apparently suitable food by the Blue whale when other species were taking it in large quantities is unexplained, and if there is a similar restriction to a limited menu in other seas the Blue whale differs therein from all the other rorquals.

On the other hand Scammon [175] over a hundred years ago said that Blue whales came into a moderate depth of water off San Quentin, Lower California, in large numbers 'attracted by the swarms of sardines and prawns with which the waters were enlivened; and the whales, when in a state of lassitude from excessive feeding, would frequently remain nearly motionless ten to twenty minutes at a time, thus giving the whaleman an excellent opportunity to shoot his bomb-lance into a vital part, causing almost instant death'. He did not, however, examine the stomach contents of the whales, so his statement must be accepted with caution. He does not describe the 'prawns' – perhaps they were lobster krill. Nor is his circumstantial anecdote good evidence that the whales were in fact feeding on anything.

The Fin whale, on the other hand, appears to be willing to eat anything suitable it can find. In the northern hemisphere it eats fishes as well as euphausian krill, especially shoaling fish such as herrings, mackerel and capelin, *Mallotus villosus*, a small salmonid fish that is found in enormous shoals off both east and west coasts of northern America. In addition in some places it takes very much smaller planktonic creatures: in the waters north of the eastern Aleutian Islands in the Pacific a copepod, *Calanus cristatus*, is one of the most important foods of the Fin whale [141], as is another species of copepod in the Gulf of Alaska.

The Sei whale is even more catholic in its tastes. It eats not only all the foods recorded as taken by the Fin whale, but gorges on lobster krill in the south Atlantic and tropical Pacific. When eating this food the stomachs of the whales are generally found to be filled either with adult *Munida* or larval *Grimothea* but not both, the adults evidently taken from surface swarms as there is no evidence that the whales feed at the bottom. In the far Antarctic [133], and doubtless elsewhere, Sei whales also eat planktonic hyperiid amphipods similar to the 'brit' eaten by the Greenland Right whale in the Arctic. The hairy fringes of the baleen of the Sei whale are particularly soft and silky so that they form an unusually fine filter. Armchair naturalists have therefore concluded that it is particularly adapted for taking minute plankton such as copepods, but this is not supported by what we know of the diet of this and other species. The Sei whale certainly does eat copepods, but it equally eats much larger food, from amphipods such as *Parathemisto gaudichaudi* which swarms at the Antarctic convergence, euphausians, and lobster krill, to small fishes such as sardines, and larger ones such as herrings. Furthermore, the much coarser baleen of Fin and Black Right whales appears in practice to be equally fit for catching copepods, so that at some times and places microscopic plankton forms the main diet of these species. It is thus not correct to assume that the different textures of the baleen in different species are correlated with different diets, though at first sight it seems an obvious inference. Even the tropical and subtropical Bryde's whale, which has short baleen plates with long coarse bristly fringes, and habitually feeds on shoaling fishes the size of herring and pilchards, and on small sharks up to a couple of feet in length, sometimes feeds on planktonic crustacea.

The Minke whale, the smallest of the rorquals, does not necessarily feed only on the smaller planktonic creatures, as is sometimes stated. It certainly does feed on plankton, especially on copepods and on krill in the Antarctic, but it is also a fish-eater in other oceans. In the north Atlantic it takes herrings, and is also known to eat capelin. It is also recorded [137] as eating cod and whiting, which implies that it feeds near the bottom in comparatively shallow seas. On the other hand the Minke whales that frequent the seas off the coast of Brazil near the edge of the continental shelf and beyond during the second half of every year are not feeding, for the stomachs of nearly all are empty [211]; their feeding grounds appear to be near the ice edge of the Antarctic. Although most northern Minke whales have a white band on the flipper whereas most of the southern ones do not, they are not different species, and there appears to be no difference in their feeding habits.

The Humpback, which comes into inshore waters more than the rorquals, eats krill in the Antarctic and lobster krill in southern temperate seas. It also eats fishes, especially capelin, in addition to *Thysanoessa* krill, in the north, but apparently does not habitually take herrings; Millais [124] states that it also eats squid.

The fish-eating rorquals, the Fin, Sei and Minke whales, in the north-west Atlantic now face a new competitor for their food. A fishery for capelin which takes half a million tons of the fish annually has been established and Sergeant [189] states that the whales 'have developed a strong tendency to come around fishing vessels and take the fish passing through the meshes of purse seines and of trawls at the surface'.

The food and feeding habits of the Gray whale are particularly interesting because they have been watched at close quarters in captivity. The Gray whale lives in coastal waters on both sides of the Pacific, feeding in the far north during summer and migrating up to 6,000 miles south to warmer waters for breeding during the winter; the herd that winters on the coast of Lower California spends the summer in the Bering Sea, whereas the one that winters on the coast of Korea summers in the Okhotsk Sea. Both herds feed during their stay in the north, but appear to eat little or nothing in their summer quarters where the young are born and the females inseminated.

In 1972 the American cetologists Carleton Ray and William Schevill studied the feeding of a yearling female Gray whale kept in a sea aquarium at San Diego, California [166]. Her keepers had trained her to eat frozen squid, of which she ate 900 kilograms daily. Dead squid are not a natural diet for any whale, but they suited this one well, for she gained about 40 kilograms in weight a day. The keepers dumped the squid into her tank frozen into 9-kilogram blocks which floated at the surface, but as they thawed the squid separated and sank to the bottom. The whale took the food from the bottom in an unexpected but apparently natural manner, since her keeper had not taught her. She rolled over about 120°, so that she was rather more than on her side, bringing her cheek parallel with the bottom. She swept over the bottom a few centimetres clear of it but did not open her jaws widely to feed; she opened the lower lip slightly and sucked in the dead squid through the side of her mouth. The food apparently was sucked in under the ends of the baleen plates by retracting the tongue and expanding the throat so that the two throat-grooves opened out. She then expelled the water between the baleen plates and swallowed the food. This feeding was selective, for when several sorts of small fish were mixed with the squid she sucked them all in but spat out the fishes and swallowed only the squid. On the other hand the stomach of a Gray whale stranded on the coast of Washington contained several gallons of rainbow smelts [69]. The captive whale always turned on her left side and fed through the lips of the left side, perhaps because her keepers trained her to take food on this side while they were teaching her to accept an unnatural diet, for the majority of Gray whales in the wild appear to be right-handed [169].

The American cetologists consider that the method of sweeping close to the bottom with one side of the mouth is the normal way of feeding for the species. It explains why the stomach contents of Gray whales examined in the north are filled with benthic, that is bottom-living, creatures. Although many kinds of prey have been found, by far the greater part of that eaten consists of about six species of amphipod crustaceans $\frac{1}{2}$ to 1 inch in length – amphipods similar to the familiar sandhoppers that swarm at the tide line of sandy beaches. These creatures live buried in the sand but when disturbed they move just off the bottom. This way of feeding also explains why Gray whales have often been seen with mud sticking to the rostrum and back or flanks – and why one of the names of the old Yankee whalers knew them by was 'mussel-grubbers'.

Further evidence that this is the normal way of feeding with the Gray whale is provided by the observations of Kasuya and Rice [96], who found that the barnacles which encrust the head of this species are absent from the lips and jaws on one side of the head, and that the baleen plates of the same side are shorter though not less in number than the opposite series. Feeding by turning on one side to take in bottom-living benthic Crustacea abrades the barnacles from the skin of the lower side, or prevents their larval stages from becoming attached, and the greater wear on the baleen plates of that side causes them to be shorter. Out of thirty-one Gray whales examined only three were left-handed, all the others fed on the right side of the mouth. Nevertheless Gray whales have been seen swimming around shoals of small fishes, and gulping mouthfuls of them by rising vertically through the shoal so that the head was brought out of the water [199].

The habit of rolling on the side when feeding is also seen in the rorquals and the Humpback. In the Fin whale it has been suggested that the habit is correlated in some way with the asymmetrical coloration of the body and baleen. The dark coloration of the upper surface is lopsided so that it reaches farther down the left side than the right, and consequently the left lower jaw is dark but the right one is white, as is the right side of the front of the rostrum. The baleen plates of both sides are dark except those of about the front third of the right side, which are white. The Fin whale when feeding rolls towards the right as it opens its mouth to engulf swarms of krill; it similarly rolls when diving after blowing. The Blue whale, too, rolls towards the side when feeding, although there is no asymmetry in the colour of the body of baleen. About 10 per cent of Sei whales show some degree of asymmetrical body colour, but it is uncertain whether this species also rolls when feeding. It is possible that it may do so when eating Antarctic krill, but in lower latitudes where it feeds on much smaller Crustacea, chiefly copepods, it is unlikely to do so. The Japanese cetologist Akito Kawamura has made experiments on the filtering efficiency of the baleen of this species, and has calculated the weight of food needed daily; he concludes that the Sei whale can collect its food more effectively by skimming than by gulping [97]. This would fit in with Millais's statement that the Sei whale does not roll when feeding as do the Blue, Fin and Humpback whales – if it is skimming and thus using its baleen like a net it would swim ahead on an even keel with the mouth partly open in a manner similar to the Right whale when skimming.

In the southern subspecies of the Minke whale some of the baleen plates are dark or partly coloured except towards the front of the rows where they are white; the coloration is asymmetrical, for the white plates are more numerous, and so extend farther back on the right side [211]. As already noted, the Minke whale includes bottom-living fish such as cod and dogfish in its diet so it is possible that this species too rolls on the side when feeding on them.

Swimming on the side when feeding from the bottom is somewhat analogous to the adaptation of some fishes to this way of feeding. The Heterosomata or 'flat fishes' lie on the bottom on the left or right side according to genus, and have evolved so far that the eye proper to the under side migrates during larval metamorphosis to a position on the upper side. The mouth, however, retains approximately its original position so that the jaws open sideways, not up and down. The advantage of this arrangement of the jaws can be verified by anyone who tries to pick a small object up from a table with a pair of forceps. It may be that even when feeding on concentrated but stratified shoals of plank-tonic krill the whalebone whales take a larger mouthful at a gulp by rolling on their sides than they would if swimming on an even keel. The evolution of the flat fishes has led to profound changes of structure, whereas that of the whales has brought only an asymmetry in the coloration of the body or baleen, or both, to some species.

Although most of the mysticetes take different sorts of food at different times and places they do fall into two sections according to their method of feeding, the gulpers that take food by the mouthful and squeeze the water out at the sides of the mouth, and the skimmers that swim with the mouth open continuously filtering until enough plankton has accumulated to be swallowed. The gulpers are the Blue, Fin, Bryde's, Humpback and Minke whales and sometimes the Sei whale when it feeds on fish; the skimmers are the Greenland and Black Right whales, the Sei and Gray whales. A diagram showing the food chain from the primary producers, the plant plankton or phyto-plankton, to the whales has been devised and modified by several cetologists [126, 132]; and a further modification of it is given in Figure 10.

The biochemistry of the planktonic food chain is unusual; the fat or lipid in food chains of land animals consists of triglyceride composed of three molecules of fatty acid esterified to one molecule of glycerol, and

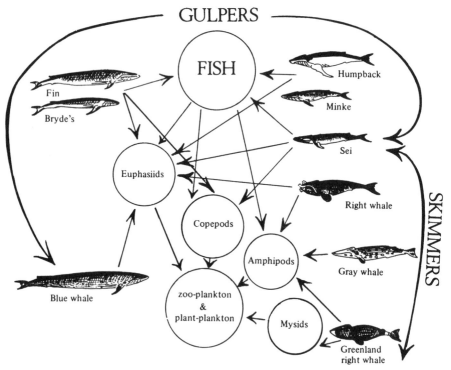

Figure 10. Diagram of the food chain of the baleen whales. The Sei whale is sometimes a gulper, sometimes a skimmer.

this is true also for the lipids in the plant plankton. On land the triglyceride is transferred to the animals that feed on plants, but in the sea the grazers that feed on the plant plankton, the copepods, the mysids and the euphausians, convert the fatty acids of the plant lipids into fatty alcohols so that their lipids are wax esters, consisting of one molecule of fatty acid esterified with one molecule of fatty alcohol. The wax ester forms an energy reserve which is necessary because the planktonic animals can feed only when the seasonal outbursts of plant plankton provide abundant grazing.

Dr J. R. Sargent points out [173] that the wax esters thus make up a large part of the total organic matter in the sea, and summarizes recent research, which concludes that the fishes and cetaceans that feed on the animal plankton use the protein in the food for reconverting the wax ester back to triglycerides by the mediation of an enzyme or enzymes. Similarly the Crustacea derive their fatty alcohol from the fatty acids of their plant food by the mediation of enzymes. The cetaceans, however, especially the odontocetes, have a high content of wax ester in their tissues, especially in the melon and the jaws. The importance of the melon lipids, in the form of wax esters, in the life of the odontocetes is discussed in Chapter 7 where the subject of echolocation is considered. It is peculiar that the lipids of the whales that feed on plankton, the mysticetes, are triglycerides whereas the wax esters are found in the odontocetes which feed on squids and fishes. Perhaps the wax esters pass through the food chain direct to the squids and thence to the odontocetes, for any lipids in fishes are triglycerides. If they do not pass unchanged through the squids, then the odontocetes must themselves produce wax esters from triglycerides by their own metabolism. Evidence that the odontocetes do in fact produce their own wax esters has been published by Varanasi and his colleagues who find that the biosynthesis is largely independent of fluctuations in fatty acid content of the lipids in the food [207].

For a hundred years marine biologists have been studying the lives of planktonic organisms, from bacteria and diatoms to larval fishes, and have amassed much information about the species used as food by the baleen whales. More work has probably gone into research on the natural history of krill, *Euphausia superba*, the most important animal of the Antarctic plankton, than on any other single species, because the once great Antarctic whaling industry depended solely upon its abundance.

Euphausians of many species live in all the oceans of the world and wherever they abound in the plankton they are preyed upon by whales. All of them are small, ranging from less than an inch to about 3 inches in length; they resemble shrimps and prawns in general appearance but are taxonomically segregated into the order Euphausiacea. This separates them from the other order, Decapoda, containing the shrimps, lobsters and crabs, with which it forms the super-order Eucarida of the class Crustacea. The larva that hatches from the egg is unlike the adult, whose character it progressively assumes at successive moults during the course of growth.

The eggs of *E. superba* hatch in the deep southward flowing current that comes to the surface on reaching the Antarctic continent and then drifts northwards as a cold surface current until it dips below the warmer waters it meets at the Antarctic convergence between latitudes 50° and 60°S. The larvae have metamorphosed by the time they come into the surface waters, where they feed on the rich plant plankton consisting mostly of diatoms; the animals breed first when two or three years old and may live to about five. Most of the krill is concentrated in a comparatively narrow belt under the ice until the end of the year when the ice breaks up leaving clear water. The whales arrive when the ice breaks up so that their food becomes accessible, and at the same time the drift from the Weddell Sea forms a northerly loop that brings the feeding zone to the South Georgia feeding grounds from October to March [109]. There is thus a fairly narrow feeding zone right round the Antarctic, covering about 16.5 million square kilometres, in which the whales graze on the the shoals of krill.

The shoals are concentrated into lens-shaped swarms, generally less than 40 metres from the surface, and often no more than 5, where they feed on the rich flora of diatoms in the layers penetrated by light. The krill swarms vary in size from less than a metre across to over 150×400 metres, the density of animals, which all swim facing the same way, increasing from the edges to the centre; swarms reach a weight of 15 kilograms or more a cubic metre. Concentrations occur particularly in regions of turbulence or upwelling, in which the plant plankton is especially rich [113].

The total volume and weight of krill in the feeding zones and the amount eaten by whales are astonishing. The Russian biologist Zenkovich has made some interesting calculations about them [215]. In the sixty-one years from 1904, when Antarctic whaling began, to 1966 when it had declined disastrously because of the scarcity of whales, 331,142 Blue whales were taken, as were 671,092 Fin whales, 145,424 Humpbacks and 87,284 Sei whales – nearly half the latter in only two of the latest seasons because of the great decline in numbers of Fin whales. The original stocks in 1904 before exploitation were about 100,000 Blue whales, 200,000 Fin, 50,000 Humpback and 75,000 Sei. The whales fill their stomachs at least four times a day, and spend about 120 days feeding in the Antarctic each season; they digest their food quickly, in three to four hours, and consequently also fatten quickly.

The Blue whale eats about one ton of krill at a meal, the Fin whale about 700 kilograms, the Humpback 500 and the Sei whale 300; the total weight of food a day is thus about four tons for the Blue whale, three tons for the Fin whale, two tons for the Humpback and one and a half tons for the Sei. In a feeding season of 120 days, therefore, a Blue whale takes some 480 tons, a Fin whale 360 tons, a Humpback 240 tons and a Sei whale 150 tons. In 1904 the Blue whales must have eaten 50 million tons a season, Fin whales 72 million tons, Humpbacks 12 million tons and Sei whales $11\frac{1}{4}$ million, so that the unexploited stocks of all species together before 1904 must have eaten nearly 150 million tons of krill, mostly *E. superba*, each season in the Antarctic.

Another Russian worker, P. A. Moiseev, also says that dense concentrations of krill south of 60°S. from December to March reach densities of 10 to 16 kilograms per cubic metre of sea water, in irregular swarms extending to several hundred metres in the upper layer down to 5 metres, rarely over 40 metres. He found that individual *E. superba* in the Scotia Sea reach a length of 60 to 70 millimetres and a weight of 1 gramme; they contain 7 per cent by weight of fat and 16 per cent of protein [128]. The nutrient salts, especially those containing nitrogen and phosphorus, and the standing crop of plant plankton nourished by them are much higher south of the Antarctic convergence than north of it [82, 174]. The great abundance of plant plankton in the southern ocean is due to the richness of the diatom flora which may make up 99 per cent of it. Dinoflagellates are also present, but they are not primary producers because they do not contain chlorophyll, and they are all phagocytes. The diatoms provide food for euphausians, and for the copepods which make up the bulk of the animal plankton at greater depths.

The late Dr N. A. Mackintosh – one of the writer's *Discovery* Expedition companions of 1924 – has ventured some opinions on what is happening to the krill now that the consumption of it by baleen whales is greatly reduced [109]. He points out that although there was severe over-fishing of the Humpback in 1910 to 1915, it was the development of pelagic whaling in the late 1920s that caused the real trouble to the stocks of whales. He calculates that over two million tons of whales were caught in the 1930s, and little less in the 1950s after the pause in whaling during the war. This was followed by a steep fall in catches in the 1960s, so that the original stocks of 1904 had been reduced by 85 to 90 per cent. The weight of whales supported by Antarctic krill has been reduced by about twenty million tons, so the presumed surplus of krill which would have fed it is not less than 30 million tons and may be as high as 300 million. We might expect to see a great increase in the numbers of other animals that feed on krill – penguins and other sea birds, Minke whales, crab-eater seals, oceanic fish and perhaps squid. There is no evidence to show that any of them except fish, birds, and the Fur seal are more plentiful than formerly.

In 1963 a pelagic gadoid fish, the Blue whiting *Micromesistius australis*, previously known from the continental shelf of Patagonia, was found in dense shoals inhabiting the Scotia Sea [123]. It is now known to form similar dense shoals feeding on krill in South Georgia waters and elsewhere. The Blue whiting is a highly palatable fish, and is now being caught in great quantities by Russian and Polish factory ships. Was it there in such great numbers before 1963? It seems inconceivable that it could have been overlooked by the whalers, and have remained unknown for some sixty years. The whalers well knew the shoals of another large fish, *Notothenia rossi*, which gathers in immense shoals to feed on krill swarms a few metres below the surface. They were in the habit of catching large numbers of them as food for all hands at the whaling stations – like most fish of the Antarctic they are edible but lack the fine flavour of the commercially marketed sea fishes of the northern hemisphere. It may well be that the unexpected arrival of Blue whiting in such great numbers was due to a population explosion of the species brought about by the presence of millions of tons of krill formerly eaten by whales but now available as food for the fish.

The surplus of krill may also have contributed to the spectacular increase in the population of the South Georgia Fur seal, once numerous but almost exterminated by sealers during the nineteenth century. In the 1920s a few Fur seals were seen from time to time, rather more in the 1930s, but by the mid-1950s the herd breeding at the north end of the island had increased to about ten thousand and by the mid-1970s has reached over a quarter of a million; in 1975 no less than ninety thousand young were born [151]. These animals eat fish and squids at some times and places, but Nigel Bonner found that at South Georgia krill formed their staple diet [22]. We cannot know whether the increase has been possible because less krill is eaten by whales so that there is more for seals, or whether there were fewer whales in the days of the former abundance of seals; it is certain, however, that there is enough krill to support a large population of seals, as well as hordes of penguins, other sea birds and fish. It will be interesting to see how the Antarctic ecology centred upon krill develops in the future, because krill is now being directly cropped by man – Russian factory ships are catching krill in bulk for processing into products for use in agriculture, and as human food.

Turning now to the odontocetes, which do not feed on plankton in bulk, we find that far less attention has been given to the animals that make up their diet. We have already noted that some of the odontocetes have numerous pointed teeth that are generally supposed to be adapted for the capture of slippery active prey such as fish and squids, but that others, particularly the ziphiids and narwhal, are almost or quite toothless although they feed mainly on the slipperiest prey, the squids. No satisfactory explanation of the apparent paradox appears to have been found.

Most of the information about the food of toothed whales comes from the examination of the contents of the stomach in whales killed in the fishery or accessible to zoologists in other ways, often by fortuitous strandings. Digestion in these animals appears to be rapid so that generally only the hardest parts of the food remain for examination. The hardest and densest bones of teleost fishes are the otoliths or ear bones, which are concerned with body orientation rather than hearing. They often have a characteristic shape so that the species of fish from which they come can be determined. Jaws and other bones of the skull can also often be identified. The parts of cephalopods that are not destroyed by digestion in whales are the horny jaws, resembling the beak of a parrot, that lie in the mouth at the centre of the circlet of arms carrying suckers. Large quantities of fish otoliths and cephalopod beaks are often found in the stomachs of odontocetes, but as they are probably the accumulation of remains derived from many meals they show only the kinds of food eaten and give no information about quantities.

The cephalopods eaten by odontocetes are wrongly called 'cuttle fish' in some books. Cuttle fish generally live on or near the bottom, unlike the squids which are pelagic or oceanic, generally rising towards the surface at night and going to greater depths by day. Many species of them shoal like fishes and make extensive seasonal migrations. Living squids are very different in appearance from the illustrations made from dead specimens in systematic and other works, in which the arms and tentacles are spread out to show their structure. In life the two long tentacles used in capturing prey are retracted out of view, and the eight short arms are held pressed together so that they appear to be a solid conical structure at the front of the head. The animals are thus so unlike the conventional illustration that at first sight they can well be mistaken for fishes. Indeed the superficial resemblance of these cephalopods to fishes in structure and behaviour is so great that it has been carried to 'the limits of convergence'. [145]

The stomachs of odontocetes taken in commercial fishery, however, often contain recently captured food so that the different kinds of prey can be identified and the relative quantities determined. Although particular prey may be preferred at different times and places, the odontocetes in general are opportunist feeders, and consequently where information is available a long list of different species makes up the diet. The dolphins and porpoises are known to feed on many species of fish and squid but in addition have been found feeding on shrimps [126], jellyfish [145] and other creatures.

The Sperm whale, as is widely known, feeds on giant squids, which however by no means form its only diet. It no doubt takes giant squids whenever it can get them but smaller species from 1 to 3 feet in length are commonly found to make up most of the stomach contents – if the soft parts have been digested small squid beaks by far outnumber the large ones. Fish of many species are also taken. Although giant squids are not so huge as popularly supposed they are nevertheless of impressive size. Measurements are misleading – a squid more than 50 feet long overall is not so formidable as it sounds, for the greater part of this length is taken up by the long extensible tentacles, and the comparatively slim body is less than 20 feet long. Some of the suckers at the end of the long tentacles of the giant squid *Architeuthis* are modified as claws, so that the skin of many Sperm whales carries scars showing as white marks on the dark background, made by the suckers and claws of the prey. The circular marks show that the chitinized hard edges of the suckers that made them can be up to 4 inches in diameter – the size of a pint mug. The scratches, like the sucker marks, usually occur on the head and round the jaws and often converge towards the mouth. They may be up to 9 or 10 feet in length and appear to have been made by the prey clinging desperately to the whale in resisting being swallowed by it. Scratches are always more numerous than sucker marks; both are uncommon on females which, being much smaller than the bulls, appear to feed on smaller cephalopods and not to attack the giants as the bulls do. In Sperm whale stomachs the partly digested remains of squids are less often found than the beaks and eye lenses which resist digestion. Prey up to 7 or 8 feet in length appears to be swallowed whole, for squid 'pens', the skeletal stiffening of the body, 6 feet long have been found intact. Giant squids are broken up before being swallowed; the soft parts from them are usually found as chunks of sucker-bearing arms, buccal masses and other fragments. The squid beaks in the stomach are sometimes numerous – six to seven hundred small ones have been found in one stomach, but large ones are generally few. Fish remains are far less frequent, but show that fish up to over 3 feet long are sometimes eaten.

The *Discovery* Expedition zoologists found that squid remains are commoner in Sperm whales caught near South Georgia than in those taken off South Africa [120]. It may be that squid are more abundant in southern waters because they feed on krill, but so little is known about the distribution, numbers, food, and the general biology of the animals that it is impossible to be precise. Squid are very active animals and avoid being caught in the usual gear used by zoologists for capturing fishes, so that much remains to be discovered about them. Although naturalists' nets catch few, their great abundance is shown by their remains in the stomachs of toothed whales, many kinds of seal, and of oceanic birds from albatrosses down to petrels.

In the 1890s Albert I, Prince of Monaco, who devoted some of the immense revenue from the gambling tables of Monte Carlo to oceanographical research, made several cruises in his steam yachts to kill Sperm whales in order to secure specimens of parts of giant squids from their stomachs. To this day no one has found a better way of collecting such specimens. A paper published in 1976 [38] remarks, 'Because Sperm whales sample larger cephalopods than other animals or man-made samplers, they give a unique view of the cephalopod fauna of any region and make an otherwise unobtainable contribution to our knowledge.'

The squid beaks also give information about the travels of the Sperm whale, for the same paper records the beaks of two species not known to occur north of 40°S. in the stomachs of Sperm whales caught off the coast of Peru in latitude 5°S., showing that the whales had recently moved north from the sub-Antarctic regions towards the Equator. The identification of squids, and the estimation of their size, from examination of the beaks is not easy, but the art has been perfected during the last decade by the researches of several zoologists.

Beale [13] states that the Sperm whale eats fishes only when in inshore waters; whether this is correct or not, they evidently form only a small part of the diet. He also says that when feeding the Sperm whale remains still and opens its mouth, letting the lower jaw hang down, and swallows the cephalopods which are attracted by the white teeth and lining of the mouth. There may be some truth in this theory which is supported by the fact that in the Mediterranean and elsewhere squids are caught on an arrangement of unbaited hooks embedded in a white-painted sinker which acts as a lure. It seems unlikely, however, that the Sperm whale catches giant squids in this way because these squids are believed to live at great depths to which so little light penetrates that the possibility of seeing the white jaw of the whale seems remote.

The Killer whale is the only odontocete that feeds on warm-blooded vertebrates as well as on fishes and squids. Stomachs of this species have been found to contain porpoises, seals and birds, sometimes several swallowed whole, but not the great number sometimes stated. Slijper [194] gives a drawing that implies that the Killer whale examined by Eschricht in 1861 contained thirteen whole porpoises and fourteen seals; if he had read Eschricht's paper with care he would not have made this blunder, for the stomach contained only the remains of these animals, some in very small fragments, showing that the whale had indeed eaten them, but not at a single meal. Killer whales, which go in small schools of about half a dozen, sometimes attack the larger whales, but the popular yarn repeated from book to book that they attack full grown baleen whales by fastening upon the tongue and biting it out seems to be based on a misapprehension. In 1725 the Hon. Paul Dudley sent an essay on the natural history of whales to the Royal Society, which was published in the *Philosophical Transactions* [50]. He gives an account of the whales and the whale fishery of the coast of New England. He describes the baleen whales and the Sperm whale, and then the Killers 'that prey upon the Whales, and often kill the young ones, for they will not venture upon an old one, unless much wounded'. He goes on,

'They go in company by Dozens, and set upon a young Whale, and will bait him like so many Bull-dogs; some will lay hold of his Tail to keep him from threshing, while others lay hold of his Head, and bite and thresh him, till the poor Creature, being thus heated, lolls out his Tongue, and then some of the Killers catch hold of his Lips, and if possible of his Tongue; and after they have killed him, they feed chiefly upon his Tongue and Head, but when he begins to putrefy, they leave him.

This essay appears to be the source of the exaggerated story, repeated uncritically for two hundred and fifty years.

Killer whales do sometimes feed on baleen whales; a school killed and partly ate a Minke whale on the coast of Vancouver Island in recent years [77]; Scammon saw a school attack and kill a young Gray whale, and Killers have been seen attacking Minke whales elsewhere. A group of seven or eight Killer whales that frequented the coast of Lower California in 1967–8 was in the habit of harassing Gray whales, and once killed a calf and ate the blubber from its lower surface. This is stated to be the first record of attacks by Killers on Gray whales during many years of close observation on the Californian coast [10]. On the other hand all whalers know that Killers sometimes bite lumps out of the dead adult whales of the larger species that are being towed to whale factories.

The White whale, which comes into coastal and estuarine waters in summer, evidently feeds at the bottom in shallow water, for flat fishes have been found in its stomach in addition to the round fishes commonly taken. Most of the smaller odontocetes, however, take their prey in mid-water rather than from the bottom even in shallow coastal waters, and oceanic species, particularly the ziphiids, must necessarily do so. The depth at which most of the ziphiids feed is not known, but the Bottle-nosed whale, *Hyperoodon*, can go to great depths when harpooned, and may also do so in normal feeding. It would thus resemble the Sperm whale, which is known to go as deep as 1,700 feet, presumably in search of giant squids. The Pygmy Sperm whale may be unlike most odontocetes because there is some evidence that it feeds at the bottom of the sea – its underhung lower jaw also suggests this possibility.

We have already noted that cetaceans in general are opportunist feeders although many species show a preference for certain kinds of food. Their migrations are correlated with the seasonal abundance of food in different places, though what signals from the environment, or what internal rhythms, send the rorquals off to the Antarctic at the right time to find the immense biomass of krill remains unknown. Day length or water temperature may well be involved for the rorquals, but ocean currents and drift are sometimes important for odontocetes, as in 1937 when an unusual invasion of the North Sea by water from farther south in the Atlantic came round the north of Scotland bringing great numbers of squid, *Todarodes sagittatus*, with equally unusual numbers of Common dolphins in pursuit of them.

One might think that it would be impossible for a cetacean to leave the water and catch its food on the land, yet such a happening has indeed been recorded. In 1964 and 1966 H. D. Hoese found Bottlenosed dolphins inhabiting a tidal creek draining a salt marsh near Boboy Sound, Georgia, where there are plenty of small fish during the summer months [86]. The dolphins entered the creek at every low tide to feed, presumably finding the fish in the muddy water by echolocation. In the autumn and spring the fish were fewer, so the dolphins hunted in pairs apparently to round up the small shoals and drive them to the edge of the water, whereupon they 'suddenly rushed up the bank together, usually both on the right side and pushing a large bow wave ahead that broke on the bank immediately before them'. At each rush the wave carried with it several fishes 3 to 4 inches long which were stranded and eaten by the dolphins when the wave broke. Usually the whole body of the dolphins came out of the water, sometimes with the tail flukes as well, and the animals quickly picked the fish off the mud and then slid back down the slippery bank. Hoese noted the remarkable mobility of the neck of the dolphins when they caught the fish 'with several surprisingly agile biting movements of the head' and added 'it is difficult to understand why such an apparently agile animal cannot capture fish in the water once the prey is congregated in a school'. This manoeuvre that the dolphins had invented was possible only for a short time each tide when enough of the banks was exposed to allow the fish to be stranded and the dolphins to slide back. It was no isolated instance, but a regular habit often seen during two years, and perhaps a local 'tradition' learnt by young ones imitating the older ones.

The cetaceans are thus one of the end results of the great productivity of the sea. The huge biomass of the whale stocks, apart from the portion removed by man in historic times, represents a point where the second law of thermodynamics breaks the metabolic chain. As each generation dies off its remains on dissolution return to the beginning of the food chain to continue the ever-turning cycle.

Chapter 4

Breeding and growth

At first sight it may seem that, as all animals are mortal, their most important activity is to find their food, so that they can stay alive as long as possible. Feeding, however, is only a means towards reproduction, which is the main purpose of life, if life can be said to have any purpose. Reproduction provides offspring to replace the parents and pass their genes on to succeeding generations.

We flatter ourselves by thinking that we as individuals are what matters, and that our germ cells are merely part of us. The reverse is true; we are merely the vehicles for housing and transmitting de-oxyribosenucleic acid (DNA) that carries the genetic code. It is not possible in the present state of knowledge to guess why the self-replicating DNA molecules should have produced such an enormous variety of living things to carry them as an endless stream for millions of years, so that they are in effect immortal. It is another question altogether whether it is possible to regard such molecules as living and thus subject to mortality.

However that may be, the cetaceans are subject to the basic patterns of mammalian reproduction in their anatomy, physiology and behaviour, with modifications to suit their particular way of life as aquatic mammals. The streamlined shape of the body must not have any structures projecting that might cause turbulence in the water, and consequently the mammary glands of the female and the genitalia of the male are carried internally and do not interrupt the contours of the body. The sex of a cetacean is thus not immediately apparent; closer inspection shows that the vulvar cleft of the female is near the anus, whereas the slit through which the penis of the male can be extruded lies some distance in front of it nearer to the navel.

The mammary glands of the female lie under the blubber, but outside the muscular wall of the abdomen, one on each side near the vulva, and the nipples lie retracted within smaller clefts at the sides. The glands are elongated ovals in shape, extending forwards and slightly outwards from the nipples; the thickness of milk-secreting glandular tissue is only a few centimetres except during lactation when it is greatly increased. The lobules of the gland in which the milk is produced are connected by small ducts to a large central duct which expands to form a sump near the nipple. Whale milk is rather thick and viscous in consistency for it contains up to 50 per cent of fat and 12 per cent of protein though not more than 2 per cent of sugar – the milk of the domestic cow contains only about 4 per cent of fat.

Female cetaceans have a bicornuate uterus, the horns joining to form a short body ending in most species with a cervix projecting into the upper end of the vagina. The uterus is supported by a fold of peritoneum, the broad ligament, with the ovaries at its outer edges near the ends of the horns, which terminate in the Fallopian tubes with fimbriated funnel-like mouths to receive the eggs liberated from the ovaries. The vagina, leading from the uterus to the vulva, bears a number of valve-like folds in its upper part, somewhat like those in the vagina of the female hippopotamus; but there appears to be no good reason for considering the presence of such folds to be correlated with the aquatic habitat.

In male cetaceans the penis lies under the blubber outside the abdominal muscles with its pointed tip inside the genital slit. It consists of a thick envelope of fibrous tissue enclosing the fused corpora cavernosa, below which the urethra runs embedded in the corpus spongiosum. At its root the corpora cavernosa separate and are attached to the pelvic bones which, though greatly reduced in size, are thus not rudimentary but serve an important function in giving them anchorage. The penis is long – it may be over 10 feet long in large rorquals – so that when withdrawn it is thrown into a sigmoid or S-shaped curve by the action of the retractor muscles that pull it in. The testes lie within the abdomen tailwards of the kidneys, for there is no external scrotum to hold them. Nevertheless in at least some odontocetes, for example the dolphin *Stenella frontalis*, there is a slight descent of the testes, for the gubernaculum at the hind pole draws them backwards into pouches of the peritoneum in the pelvic region; this phenomenon might be regarded as showing that the ancestors from which whales have evolved possessed a scrotum into which the testes descended.

In male mammals that have scrotal testes spermatogenesis is inhibited if the temperature of the testes is raised to that of the inside of the body. This effect is often given as the 'reason' for the presence of an external scrotum, but it may equally be that the testes show this phenomenon as a result of being in a scrotum. The condition in whales and a few terrestrial mammals such as elephants, sloths and some others, shows that a low temperature is not inherently necessary for spermatogenesis to occur. Although the internal position of the testes certainly aids streamling in whales, external testes in a large scrotum are found in the bulls of otariids, the sea-lions and fur seals, whose bodies also are streamlined in shape. On the other hand the testes of phocid seals are not contained in a scrotum but are inguinal in position; this is perhaps correlated with their mode of progression on land, for their limbs do not support the body clear of the ground whereas those of the otariids do.

Apart from a prostate surrounding the urethra at the point where the vasa deferentia join it, male cetaceans do not possess accessory genital glands such as vesiculae seminales.

The breeding season in cetaceans is not, as far as present knowledge goes, a sharply marked occurrence, for births may occur over a considerable period; there is, however, generally a peak in frequency of births which represents the centre of reproductive activity. The seasonal timing of the peak appears to vary in different parts of the range of some species, as in the Bottle-nosed dolphin [55], a species commonly kept in captivity so that its breeding behaviour has been frequently observed. For the greater part, however, knowledge of the breeding season has come from the examination of dead animals, whether pregnant, lactating or reproductively inactive females. That being the case, it is not surprising that most of the work on the subject has been carried out on the whales brought in for processing to commercial shore whaling stations or floating factories – the rorquals and the Sperm whales. Knowledge of the reproductive cycle in most of the smaller odontocetes is more fragmentary, and is derived from animals examined as opportunity occurred.

The accumulated information about the presence and the size of foetuses in pregnant whales, and the dates when they were found, show that in the southern oceans most of the young are born in the winter when the whales have migrated to comparatively warm temperate and subtropical waters. Examination of the internal organs of female whales, especially the ovaries, gives much more information about the sexual cycle and life history of whales than merely the time and duration of the breeding season.

The ovaries of all mammals contain great numbers of germ cells at birth; the cells are differentiated at an early stage in the life of the embryo and are not added to subsequently. When the animal reaches sexual maturity the egg cells ripen in succession and each is surrounded by a layer of nutritive cells that secrete a fluid in which the cell is suspended, the whole forming the ovarian follicle. The mature follicle or follicles – for in animals that produce litters more than one ripens at the same time – bulge from the surface of the ovary and finally burst, each liberating its egg cell into the funnel at the end of the fallopian tube, down which it is wafted by the action of the cilia of its lining to be fertilized by fusion with a spermatozoan received from a male. It then passes into the uterus where it becomes attached to the lining, and develops into a foetus with its enveloping membranes and the foetal part of the placenta. After the follicle bursts its lining cells enlarge to fill the space, and take on a new function, secreting the hormone progesterone that stimulates the lining of the uterus to activity for the reception and nourishment of the fertilized egg, and helps to maintain the pregnancy. The colour of this body is usually yellow, hence its name, corpus luteum. During pregnancy in some mammals, and after the birth of the young in others, the corpora lutea stop secreting and degenerate; in some species they are absorbed so that sooner or later they disappear, but in others the remains persist for some time. Where this happens the degenerated cells are replaced with fibrous and connective tissue forming a body called the corpus albicans, which gradually shrinks in size and is generally completely absorbed after some time. A scar often persists on the surface of the ovary at the spot where the follicle burst.

Whales are unusual in that the corpora albicantia remain visible for a very long time; possibly they never disappear completely. It is therefore possible, by examination of the ovary, to find out how many times the animal has ovulated. As whales bear only one young at a birth it would seem that the number of corpora albicantia would show the number of young that the animal has produced. The matter is not, however, as simple as that. When a female whale comes into heat, or oestrus, under the influence of the sex hormones, the egg cell released from the ovary may for various reasons not be fertilized. In the absence of fertilization the corpus luteum rapidly degenerates into a corpus albicans, and another period of oestrus, with another ovulation, follows. In this way several corpora albicantia may accumulate before a successful fertilization and pregnancy occur. The corpus luteum of pregnancy grows much larger than those of sterile ovulations, and eventually produces a large corpus albicans. Corpora albicantia of different ages differ in size and appearance, so it is possible to distinguish the separate batches, if more than one follicle ovulated, and thus trace the reproductive history of the animal from which the ovary comes. In practice it is necessary to cut the ovary into slices for examination to ensure that nothing is missed; the ovaries of large rorquals may be over 2 feet long and weigh 20 to 30 pounds or more, so that the zoologists of the *Discovery* Investigations used a commercial bacon-slicing machine to expedite the work.

The ovaries of mysticetes are not nearly so compact as they are in most mammals, and consequently the ripe follicles and corpora lutea often project so far from the surface that they are attached only by attenuated tissue like a stalk. There is, however, an uncertainty in estimating the age of mysticetes from the number of corpora albicantia because their number may be increased by the presence of accessory corpora. Accessory corpora, described below in considering the ovaries

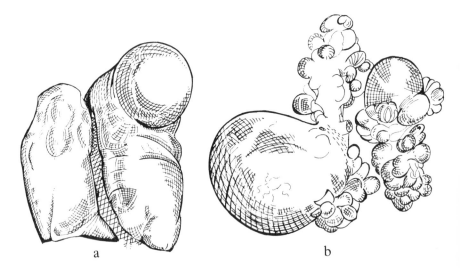

a   b

Figure 11. The ovaries of a Sperm whale (a) and of a Blue whale (b). Those of the Sperm whale are compact and show one large corpus luteum; those of the Blue whale are diffuse and show one large corpus luteum and several smaller ones and corpora albicantia.

of the odontocetes, are formed during pregnancy and do not represent fertile ovulations. The ovaries of mysticetes may contain from 1.5 to 3.7 per cent of accessory corpora, and if they are counted as representing pregnancies a false estimate of the whale's age is obtained.

Examination of the ovaries, together with observations on the growth of the foetus, the time of birth, and length of lactation, thus gives an indication of the age of the whale examined. The large rorquals bear a calf every other year, because gestation lasts nearly a year and lactation continues for over six months after the birth, so that if the age of sexual maturity at which the first successful ovulation resulting in pregnancy occurs is known, the approximate age can be calculated [110].

Useful as these methods are in tracing the life history of the rorquals, they cannot give results as precise as could be wished – they are never more than approximate. Zoologists have therefore sought for other ways of obtaining more exact information on the ages and life span of whales. The surface of the baleen is not smooth but is covered with a series of rather irregular ridges running across each plate. The late Professor Ruud, the Norwegian cetologist, noticed that the ridges and the grooves between them are not uniform in size, and consequently indicate periods of rapid and slow growth [171]. The free ends of the plates are continually worn away by friction during feeding, so that the plates grow continuously to replace the part lost in wear. He suggested that the growth takes place at different rates depending on the seasonal stresses on the whale's metabolism – when food is abundant it is quick but during the strain of migration or breeding it is slower. He was able to show that there are successive annual peaks of growth which show the age of any particular plate of baleen. This method of ageing is unfortunately very limited because owing to the loss by wear at the functional end of the plates only about four years' growth is shown for each plate, and thus, except in young whales, no information on the age of the animal is given.

Another and much more reliable method has since been devised by the English cetologist P. E. Purves [158]. This, like Ruud's baleen ridges, depends on the counting of annual growth rings, but in an internal structure not subject to loss by wear. The ear meatus of the baleen whales – the tube leading from the surface of the body to the ear drum – has a minute hole at the outside leading to a short narrow tube that runs inwards to end as a solid strand of tissue. Farther in, however, the strand appears to open up again to form a tube increasing in diameter towards the drum. Part of the drum, unlike the nearly flat drum-head of other mammals, is ligamentous and part is drawn out into a conical tube, like the finger of a glove, projecting into the inner part of the meatus. A solid narrow tapering structure that may be up to a yard long extends from the glove finger to the closed outer part of the meatus, originally incorrectly called the 'wax plug'. Early investigators thought that the plug filled the cavity of the meatus; they were misled through the difficulty of dissecting a delicate structure out from a cumbersome

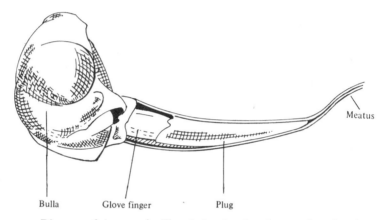

Bulla   Glove finger   Plug

Figure 12. Diagram of the ear of a Fin whale, showing the ear plug, the glove finger, the auditory bulla and the closed auditory meatus.

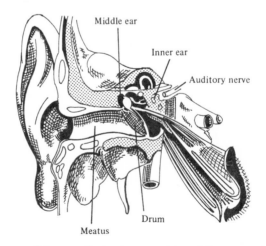

Figure 13. Diagram of the ear of a land mammal (man) for comparison with Figure 12.

mass of bone and soft tissues. The careful work of Purves showed that the plug is, in fact, the epithelial lining of the meatus, the cornified cells of which cannot be shed and discarded through the meatus as in other mammals. They consequently accumulate to form a thick horny substance surrounding rather than filling the tube of the meatus. The site of the tube thus runs down the centre of the plug but is completely obliterated.

The glove finger at the inner end also sheds its cornified cells to form a conical core in the base of the plug; by its growth it also obliterates the part of the tube that originally separated it from the plug, so that it becomes fused with it. The core differs from the rest of the plug because the glove finger not only sheds its dead cells but also secretes wax, but the shedding and secretion take place alternately so that the core is laminated, with layers of less dense waxy cells sandwiched between layers of more horny cells. The horny cells are shed during periods of growth on the feeding grounds, so that each horny layer, which is light in colour and contrasts with the darker, less dense layer, indicates a feeding migration. Purves, however, expressed the opinion that the quiescent periods represented by the light layers occurred during the migrations north and south, and thus each complete layer or lamina represents about one year of the animal's life. The distance between the layers decreases from the outside in, corresponding with the decrease in growth of the whale as it ages.

The succession of layers thus gives a record of the whale's life; the end of the first corresponds to the birth of the animal, the second, which is wide, represents the period of rapid growth during suckling, and the third, which is narrower, corresponds to the first season's feeding on Antarctic krill; thereafter the layers become progressively narrower as the rate of growth gets slower. Purves's first study was made on the ear plug of Blue whales, and he later followed this with a study of the ear plug laminations in relation to the age composition of a population of Fin whales [161]. Later work by others has confirmed Purves's conclusions and shown that similar layers are laid down even in the Bryde's whale, which has no marked migrations [36]. Perhaps the most conclusive results come from Humpbacks that were marked as calves so that when subsequently captured their ages were known; in these the number of laminations corresponded to the known age.

But even this method of ascertaining the age of whales is not conclusive. At first it was thought that two laminations represent one year

83

of life, but some marked Fin whales caught in Japanese waters showed that at least the later layers had been produced at a slower rate, approximating to only one a year. Thus it is not certain whether one or two laminations represent a year of age; however, it is probable that the rate of production of laminae varies and may decrease with increasing age.

The accumulated results of many workers using these methods, together with observations on the incidence of pregnancy and the growth of foetuses, gives a now fairly well-established picture of the life histories of the mysticetes. The gestation period varies from ten to twelve months according to species, and is about eleven months in the larger rorquals. Lactation lasts for a period of from five to seven months, but is much longer in the Humpback in which it reaches about ten months or more. In the Greenland Right whale the accounts of the old whalers suggest that lactation continues for about a year. The age at which sexual maturity is reached also varies between species and between individuals of the same species. In the larger rorquals and the Humpback it may occur at the age of a little under five years or as much as over seven years; in the Sei and Minke whales the onset is earlier and takes place at about two years of age. The latest researches, however, show that these age estimates are too low, and that the correct figures are probably several years later. Growth of the body continues for some time after sexual maturity is reached and does not stop until at least the age of ten years in the larger rorquals and Humpback; in the Blue whale it does not stop until thirteen years of age and in the Fin whale until fifteen or sometimes even twenty years or more.

The total life span is not so great as might be inferred from the great size of the rorquals. It is impossible to do more than estimate the possible longevity because even the oldest whales examined might have lived for some years longer had they not been killed. Analysis of the age composition of the commercial catch, does, however, give a fairly reliable guide. It also showed in the later years of the Antarctic whaling industry that the average age of the animals became less in succeeding seasons because the population was being over-fished so that few whales reached their potential life span. The average estimated length of life for the rorquals and the Humpback ranges from twenty-five to about thirty years. Whales older than the average have been recorded, but as in other mammals, a few exceptionally aged individuals, though rare, are to be expected.

84

The acquisition of such a comprehensive knowledge of the life of the mysticetes has been possible only by the examination of great numbers of whales in the commercial catch over a long period of years in many parts of the world. It is ironic that the means for studying the subject have been provided only at the expense of a drastic reduction in the whale populations; if the knowledge gained had been available when large scale pelagic whaling began and had been made the basis of a policy of exploitation, the whale populations would now probably be undamaged. By contrast, our knowledge of the life histories of cetaceans that are not, or have not been, exploited is fragmentary and incomplete.

Although so much is known about the vital statistics of the mysticetes, much less is known about their lives as they lead them. The breeding season extends over several months, which means that the females do not all come into oestrus at approximately the same time, and individuals vary greatly. The virgin cows are generally the first to come on heat, those that have borne calves following later, but there is no exact rule, and there is much variation. The time of the breeding season also differs in different parts of the world, for example that of the Fin whales feeding in the Antarctic extends from April to August, when the animals are in the northern part of their range, whereas in the north Pacific the season is at its peak in November. There are, however, many exceptions, for Blue whales have been seen coupling on the South Georgia feeding grounds in February.

The difficulty of seeing what whales are doing underwater and of approaching close enough for observation leaves us in ignorance of anything but the sketchiest knowledge of how the rorquals copulate. Any records are bound to be the result of fortuitous sightings and are often unavoidably imprecise. The difficulty is increased by the fact that the act of coupling is brief and lasts less than a minute, probably only a few seconds, as it does in some terrestrial animals, notably domestic cattle. It appears to be true that coupling is preceded by a considerable amount of preliminary 'play', behaviour that stimulates the partners, and particularly the female to receive the male. In land mammals the male is generally attracted to a female in oestrus by the scent of the specific secretions she then produces. In the whales some other kind of signal must serve to notify the male that the female is receptive because they are unable to use olfactory perception underwater if, indeed, they have any. It is probable that the behaviour of the female is altered, or

85

she may actively seek the male – or perhaps she gives some acoustic signal.

Among the rorquals the act of coupling appears to take place at the surface. On the South Georgia whaling grounds I once saw some Blue whales making a great commotion at the surface, splashing and churning up the water. I turned to the gunner beside me and exclaimed, 'Whatever are they doing?' His reply was a single four-letter word that showed we were having one of those rare opportunities of seeing coupling taking place, but the details of what was happening were not revealed. Similar observations reported from time to time show that the rorquals generally couple side by side at or near the surface, to the accompaniment of much stimulatory play.

The Humpback is notorious for the spectacular gambols in which it indulges both in and out of the breeding season. The whales roll over at the surface, striking the water with the flipper, raise the flukes out and crash them down raising clouds of spray, and breach – jump partly or completely clear of the water, often to one side rather than on an even keel. When they breach they do not rise into the air and return to the water in a graceful curve, as dolphins do, but fall back on their sides or bellies with a tremendous splash. As a cover for ignorance of the reason for this behaviour it is usually termed 'play' – which may be correct for all anyone knows, but is no more than guesswork. Some cetologists have suggested that this peculiar behaviour is an attempt by the whales to rid themselves of parasitic crustacean whale lice and barnacles that cling to the skin; if that is the reason the antics are singularly unsuccessful, for the bodies of Humpbacks are always heavily infested with these apparently irritating parasites.

These activities certainly take place at high intensity during the breeding season; Scammon [175] says that at that time,

their caresses are of the most amusing and novel character, and these performances have doubtless given rise to the fabulous tales of the sword-fish and thrasher attacking whales. When lying by the side of each other, the megapteras frequently administer alternate blows with their long fins, which love-pats may, on a still day, be heard at a distance of miles. They also rub each other with these same huge and flexible arms, rolling occasionally from side to side, and indulging in other gambols which can easier be imagined than described.

Side-by-side coupling, as in the rorquals, is evidently normal with Humpbacks, but Japanese cetologists have described a happening when a different manoeuvre was used [135]. Two Humpbacks were seen performing the usual splash-about antics at the surface, after which they separated, and dived and swam towards each other at high speed, then rose to the surface vertically, clasped belly to belly, so that the front part of their bodies as far back as the flippers emerged above the surface. They then fell back with a great splash, and repeated the performance several times. Similar reports have been made, though in less detail, by a number of observers from time to time, so this variant of mating behaviour is evidently often adopted.

Although Humpbacks appear to be so clumsy and ponderous in their gambols, and the shape of their bodies – short and tubby with disproportionately long and flexible flippers – is so different from the graceful slender shape, despite the great size, of the rorquals, they are equally nimble when swimming under water. Scammon well describes this when he tells how 'we have observed them just "under the rim of the water" (as the whalemen used to say), alternately turning from side to side, or deviating in their course with as little apparent effort, and as gracefully, as a swallow on the wing'.

The mating of the Gray whale appears to be similar to that of the rorquals. For the greater part it takes place in the bays of Lower California and nearby areas, but has also been seen farther north during the migration towards the Arctic feeding grounds. Gilmore, the American cetologist, considers that females with newly born calves are not receptive to the males (but see p. 91), and that, apart from the virgin cows, the females that are inseminated are those that have borne and weaned a calf in the previous season [69]. On the other hand another American observer, Houck [88] saw what appeared to be a mating during the northern migration in March 1958 off the mouth of the Mad River towards the northern boundary of California. He watched a cow with a calf followed by a male, and when the male came alongside the female,

the migration was temporarily halted and they remained in the general area engaged in a type of play. Several times both large whales surfaced together. Both sexes occasionally surfaced separately, rolled over on their sides and extended one of the flippers into the air. Sometimes they made a complete revolution. On three separate 'rolls' the penis of the male was plainly visible. The penis as observed was large, light or flesh coloured, and erected into approximately a half circle.

87

'The erected penis and the general behaviour' of the whales led him to believe he was watching a mating. The 'play' lasted about twenty-five minutes, after which the whales resumed their migration to the north.

Up to about 40 per cent of immature or virgin female Fin whales have a cord of tissue, about 10 centimetres long and 1 centimetre in diameter, crossing the entry into the vagina from front to back. It was thought to be a very unusual structure by Mackintosh and Wheeler [110] who first observed it and called it the 'vaginal band'. They appear, however, not to have known that similar phenomena are common in female mammals. During embryonic development the lower end of the vagina is closed by a solid mass of cells which makes contact with the cells of the ingrowing cloacal membrane, part of which differentiates later as the urinogenital membrane. Before or after birth the cell mass splits to make a central passage, but its former position can nearly always be identified by at least a low ridge of tissue, in some species by a definite fold partly obstructing the passage. In some mammals the spontaneous perforation does not take place until the first oestrus, and in others, such as the mole, the passage is again obstructed by fusion or healing of the edges after the birth of the young. In some mammals the breakdown of the cells is occasionally incomplete so that two small passages are formed on each side of a central remnant. This is evidently the way in which the vaginal band of Fin whales is produced. Mackintosh and Wheeler found that the outer two-thirds of the surface of the band is covered with 'epidermis' and the inner surface with fine papillae; this is to be expected because the inner part is homologous with the vaginal epithelium, the outer part with that of the urinogenital sinus. The band is not always broken when the female receives the male at her first oestrus but is ruptured at the first birth, and thereafter it may sometimes be recognized as a tag on the wall of the vaginal entrance. Although the band was found in over a third of Fin whales examined, it was present in only about a quarter of Blue whales; on the other hand it is nearly as frequent in the Humpback [117] and the Sei whale [120] as in the Fin whale. Among the odontocetes the vaginal band has been recorded only in the Dall porpoise [131].

The growth of the foetus of the mysticete is the most rapid and impressive of any of the mammals. In the Blue whale, for example, the fertilized egg cell is less than half a millimetre in diameter, so smaller than a pinhead, yet in eleven months it grows into a young whale with a length of 25 feet and a weight of about 2 tons. When the fertilized egg reaches the uterus after travelling down the Fallopian tube it becomes embedded, as we have seen, in the lining of the uterus which proliferates to receive it under the influence of the hormone from the corpus luteum. It there makes a closer contact with the maternal tissues by the development of the placenta, which is diffuse and epitheliochorial, and is connected to it by the umbilical cord. The details of the embryological development, and of the placenta and the membranes which surround the foetus, are beyond the scope of this book, but several general points should be noticed. Growth of the embryo is at first slow, and in the first few weeks the embryo is like that of any other mammal at the same stage of development, curled head to tail and with hind and fore limb buds. When the embryo is about an inch long the buds of the hind limbs are resorbed and disappear, and by the time it is a foot long at about three months, the flippers and tail flukes are developed and the general form of the body is typically cetacean. Thereafter growth is increasingly rapid, the throat grooves and the pigmentation gradually appear, and the last three months of foetal life are mainly devoted to increase in overall size and weight.

The foetus of the mysticetes, unlike that of the odontocetes, may be held in either the right or left horn of the uterus, and at first lies doubled up with the snout and tail facing the opening of the uterus. In most large mammals the young is normally born head first, and when by mischance the hind end comes first the breech presentation may cause difficulty in giving birth. In the mysticetes the process of birth is probably of short duration, but we have little information about it. It has seldom been seen; on the only two occasions when it has been carefully observed the foetus came head first. In the odontocetes, however, the breech presentation, or rather tail presentation, is normal, and the young is born tail first. In the odontocetes birth has often been observed in captivity. The first stage, when the tail projects from the vulva, appears to last for some time, maybe several hours, but the rest of the birth occurs quickly and the umbilical cord is broken when the foetus is born. The placenta is detached from the lining of the uterus and is expelled as the afterbirth generally within minutes, or at most an hour or so.

The shock to the young whale on being born must be great, when, after floating in the warm bath of the amniotic fluid in its mother's uterus for eleven months, it suddenly finds itself swimming in the cold water of the sea. In land mammals the stopping of circulation through the

89

placenta, when the umbilical cord is broken, brings a rise in the carbon dioxide concentration in the blood which reacts on the respiratory centre in the brain of the young and, with the stimulus of the drop in temperature of the surroundings, starts the animal breathing. Were this process to occur in whales the young whale would probably be drowned, as it has neutral buoyancy before its lungs are filled with air; the reaction of the mother saves it, for she nudges it up to the surface so that its blow-hole emerges and it can take its first breath. It might seem that this is the most dangerous moment in the life of a whale, but it is not so critical as it appears because the young whale, like its mother, can refrain from breathing longer than land mammals, and consequently the need to take the first breath is not so urgent.

Soon after the birth the young whale has its first feed: the mother positions herself so that it is guided towards a nipple which protrudes from the groove in which it normally lies withdrawn. The baby grasps the nipple by the tip of the jaws, and probably envelopes it by raising the sides of the tongue to wrap round it as do the young of other mammals. In the Sperm whale, however, it is said that owing to the length and narrowness of the lower jaw, the calf takes the nipple into the corner of its mouth, an allegation that needs confirmation by observation. But, as suckling occurs underwater, there is not time between breaths for a prolonged period of sucking, so the milk in the cistern of the mother's mammary gland is ejected into the baby's mouth. The muscles of the gland squeeze out a stream of milk, probably in response to the stimulus of the baby mouthing the nipple, and all the baby has to do is gulp it down. In mysticetes as the baleen of the baby grows the part at the tip of the upper jaw remains soft and short so that the young whale can grasp the nipple until the time comes for weaning.

The growth of the young whale is rapid, for the milk is rich and concentrated. A young Blue whale is said to double its birth weight in seven days – a human baby takes 120 days or more. Fin and Blue whales suckle their young, as we have seen, for only some five to seven months, but Humpbacks suckle for about ten months, and Gray whales and Right whales for about twelve. At weaning a young Blue whale may be over 50 feet long, a Fin whale nearly 40, and other species in proportion, so the amount of nourishment taken from the mother is a heavy tax on her metabolism; the large mysticetes produce as much as 600 litres (over 130 gallons) of milk a day. Throughout the nursing period, and especially in its earlier part, the young whale keeps close to its mother, swimming below, above or alongside her almost in contact. It is only when the young one starts taking some krill on the rich feeding grounds that it begins to stray far from her as the process of weaning goes on. The bond between mother and young is necessarily strong, for a young whale lost in the immensity of the ocean would have little chance of finding its parent. The mother is equally solicitous for her calf, and thus falls an easy prey to whalers if they attack it, for she is reluctant to leave even when it has been killed.

In many land mammals the female comes into oestrus soon after the birth of the young so that she becomes pregnant before she has weaned the first offspring, but in the rorquals the post-partum oestrus is not universal though it sometimes occurs, and most of the females do not conceive until after the end of lactation, so that they bear calves only once every two years. In the Humpback, however, the phenomenon is more common and consequently the rate of calf production averages about two every three years. In the Gray whale, too, contrary to Gilmore's opinion, post-partum oestrus is common [89] so that a calf is borne by most females every year. Mating has often been seen in the lagoons of Lower California after the birth of the calves, and also at the beginning of the northern migration when the female was accompanied by a young calf, as already noted.

For several years after reaching independent life the young whale devotes its activities to feeding. Rorquals attain puberty, or more strictly sexual maturity, since they do not grow pubic hair, at ages ranging from at least five years and probably more in Blue and Fin whales, a little less in Humpbacks, and less still, some two to three years, in Sei and Minke whales. The ovum of the first ovulation appears often to miss being fertilized, but the majority of the subsequent ovulations result in pregnancy. Growth of the body continues for some years after the animals reach sexual maturity; Blue and Fin whales may add another 10 feet or more to their length before growth is complete. Although the great whales live to ages less than might be expected from their size – other large mammals such as the elephant can reach three score and ten – nothing is known of the natural causes of death; it seems improbable that all senile whales succumb to the attacks of Killer whales, the only animals known sometimes to prey upon them. Whales are subject to infestations by parasites and suffer from various pathological conditions, as described in Chapter 9, some of which may be fatal to aged whales.

Turning now to the odontocetes we find that there is much less precise information about their breeding and growth than for the mysticetes, except for the Sperm whale. The reason for this is simply that apart from the Sperm whale, few of them are the subject of commercial fisheries, so a large number of their corpses has not been available for scientific examination. In a few places, notably in Japan, commercial fisheries on a small scale have produced sufficient numbers of a few species of the smaller odontocetes to enable zoologists to work out the life cycle of such species as the Striped dolphin [95], the Spotted dolphin [93] and the Pilot whale [188]. For the greater part we have to rely on the examination of stranded specimens that happen to come to the attention of a zoologist, occasional field observations of living animals, and observations on dolphins in captivity.

In female Sperm whales and other odontocetes the ovaries are much more compact than those of the mysticetes, and more closely resemble the normal appearance of those of land mammals. The surface of the ovary in the Sperm whale is smooth so that the follicles and corpora albicantia do not project far, and even an active corpus luteum of pregnancy, although causing a prominent swelling, is not constricted off by a narrow neck as it often is in the mysticetes (fig. 11). The corpora albicantia are, however, visible in the substance of the ovary when it is cut, and because they persist as visible remnants throughout life their number shows the number of ovulations that have taken place though they may not all have resulted in pregnancy [118]. As the Sperm whale is cosmopolitan in its distribution and is found in all the warmer and temperate seas of the world its breeding season varies geographically. The adult males are capable of reproduction at all times of the year, for although the tubules of the testes in some specimens show less signs of activity than those of others, they have always been found to contain plenty of spermatozoa. Any seasonal rhythm in breeding must therefore depend upon the physiology of the females. The penis of the male when extruded is even larger than that of the mysticetes and has been an object of wonder and admiration to mankind ever since illustrations of stranded whales were first made centuries ago, as we have already seen in Chapter 1.

Sperm whales are generally found in schools, except in high latitudes where solitary bulls are the rule, the schools containing up to a dozen or so females with a proprietor bull and often some smaller bulls in attendance. Little is known about the behaviour of the bulls towards each other; there is no evidence in the way of scars on the skin, such as are found in smaller odontocetes, that there is any fighting in rivalry for the females, though the damage to the lower jaw sometimes seen in bulls could be the result of hostile encounters. It is probable that there is some sort of stimulatory 'courtship' behaviour when the cow is receptive, and that copulation takes place near the surface in a horizontal position, but even today we still have few precise observations on these matters.

Gestation is much longer in the Sperm whale than in the mysticetes, being sixteen months from conception to birth. The young are generally born in the summer months of the hemisphere in which the animals live, but the seasons for breeding and birth spread over several months, so that the periods are not clear-cut. Lactation lasts about a year, and sexual maturity is not reached until the age of about nine years in cows and less than nine in bulls. The interval between births for the cows is about three years, so the reproductive rate is not high. This may be balanced by a comparatively long reproductive life, for physical maturity, the age at which growth is complete, is not reached before the age of about twenty-eight in cows and thirty-five in bulls.

These precise ages are known because it is somewhat easier to find the age of odontocetes than of mysticetes. The dentine of the teeth in odontocetes, as in some other mammals, is laid down in layers so that if a tooth is cut across the layers appear as rings, like the concentric rings in the trunk of a tree. The teeth go on growing throughout life, for they retain open pulp cavities in which new dentine is deposited. Even when the tip of the tooth wears down with use the basal part of the tooth still retains its rings. The growth rate of the teeth is not constant but fluctuates so that alternate layers of denser and less dense dentine are produced. If a tooth is sectioned and the resulting surface is suitably polished, a count of the layers shows the age of the animal from which it comes. In most of the odontocetes from which teeth have been examined in this way it is assumed that each ring represents one year of age, but in a few, such as the Sperm whale, there has been some doubt whether one or two rings are laid down annually. The latest researches, however, conclude that the rings are annual, and this view is now generally accepted [19]. The matter can only be resolved by the examination of the teeth from animals of known age, not an easy matter apart from dolphins born in captivity. It can only be done by tagging young animals soon after birth and hoping to recover them at some future date, a method which would entail an enormous amount of work in tagging a sufficient number to make any recoveries probable. But it is not so easy as it may seem to read and interpret the rings, especially in the smaller species. Furthermore the cause of the rings is not known; it is probably not due to seasonal abundance of food, for captive dolphins that are well fed daily equally show rings in their teeth.

Examination of the ovaries of Sperm whales shows that a new pregnancy starts a little before the end of lactation, and that each pregnancy is represented by about four corpora albicantia, indicating that four ovulations occur for each successful pregnancy. The conclusion drawn from the first investigation [118] into these matters was that the female Sperm whale is polyoestrous, meaning that at each breeding season she experiences several short oestrous cycles in quick succession, only the last of which results in pregnancy. There is, however, a more probable explanation that was unknown at the time; it is more likely that the female comes into oestrus and is inseminated and becomes pregnant at the first ovulation. The other ovulations probably occur at intervals during the long pregnancy and the resulting corpora lutea are in fact 'accessory corpora lutea' producing the luteal hormone progesterone that acts upon the uterus to promote the maintenance of the pregnancy. Accessory corpora lutea may be produced without ovulation taking place; the ripe follicle may not burst and shed the egg cell, but instead the egg cell may die and be absorbed while the cells lining the follicle enlarge to form a functioning corpus luteum. This phenomenon is known in some land mammals, such as the mare and the female elephant.

The probability that the first egg ovulated at oestrus is fertilized, and that the extra corpora lutea are accessory is also pointed to by the behaviour of male odontocetes. The males appear to be highly sexed so that it is unlikely that any receptive female is not at once inseminated. Little has been observed of sexual behaviour of male whales in the wild, but captive male dolphins often show strong sexual inclinations. They not only copulate with captive females when they are on heat, but try to do so with females of other species, and with other animals confined in their tanks such as turtles and large fish, and even their own human keepers who enter the tanks in diving suits to feed them. In addition they sometimes stimulate themselves with inanimate objects and tickle their genitals against bits of wood, rope or anything else that takes their fancy. Similar behaviour has also been seen in the wild; on several

94

occasions male dolphins have come close inshore or entered harbours, showing no fear of man, and engaged in similar antics. In 1973 and 1974 such a dolphin haunted the harbour at Castletown in the Isle of Man, playing among the anchored craft, and was particularly interested in inflatable rubber dinghies, following them closely in its attempts to copulate with them, greatly to the panic of the human occupants who feared it would capsize them. Later it, or a similar neurotic dolphin, appeared off the north coasts of Devon and Cornwall, coming close to small boats and similarly trying to pair with them. The same animal, or another, subsequently turned up around the Scilly Islands and then moved to the south coast of Cornwall, where its behaviour was similar. One can only suppose that these creatures were suffering from some form of neurosis, for unlike some captive animals they had access to all the females of their species in the sea if they cared to look for them, yet they had this strange obsession with man and his artefacts.

It is not possible to give a complete account of the reproductive natural history of the large number of species of the smaller dolphins and porpoises, or of the ziphiids. There is not at present any large commercial fishery for any of them on a scale similar to that recently applying to the large whales, and consequently their dead bodies are not available for study by zoologists, who have been able to learn so much about the species subject to exploitation. The examination of such specimens as come to hand from time to time has led to the accumulation of much information which is, however, fragmentary so that some points are known for certain species, others for different ones. This knowledge is increasing steadily, but it will be long before the complete story is known for any one species. An excellent review of what is known, together with much original matter, has been published by Professor R. J. Harrison and his colleagues [80].

In general we know that in the smaller odontocetes the gestation lasts about a year or up to sixteen months in the Pilot whale, and that lactation ranges from eight to sixteen months or more. As with the larger whales the smaller odontocetes show considerable maternal care of the young. A young calf also stimulates maternal and protective actions in other females which have no calf at their side – sometimes they appear to co-operate with the mother in nudging it to the surface at its birth and swimming in close company, at others the mother appears to repulse their attentions by placing herself between her calf and the would-be assistants. This reaction to a calf is sometimes misdirected to other objects, for dolphins or porpoises of several species have sometimes come to the assistance of human swimmers in difficulties and helped to support them to the surface. Such actions have, as might be expected, been recorded in romantic and anthropomorphic form by people who see more in the occurrence than is justified by the facts.

The sexual relations of the mysticetes appear to consist largely of mutual stimulation and display, with much splashing and threshing about at or near the surface. Whether the bulls are the proprietors of small schools of cows or are entirely promiscuous is unknown; on the other hand the bulls of some kinds of odontocetes certainly contend with each other, presumably for the possession of the females. Unlike the mysticetes they have teeth and not only can bite, but do. Scars on the skin showing the marks of teeth are sometimes present in some species, and always present in others. These scars can readily be distinguished from those made by other causes, such as the bites of sharks, the suckers and claws of squids, and the former presence of skin parasites. Observations and illustrations of the scarring in many species have been published by the New Zealand cetologist C. McCann [122]. He points out that as most of the odontocetes live in schools there is ample opportunity for the young males to engage in 'play' or sparring in incipient rivalry, but that the bites inflicted during play are slight and of no consequence. He maintains that when they become adult the bulls' sparring turns to serious fighting for mastery and possession of a school of females; certainly the scars show that biting has been more than slight, but they cannot indicate why they were inflicted.

A more gentle nibbling resulting in no damage characterizes the courtship of those smaller cetaceans that have been observed in captivity. A male accompanies a receptive female, the two animals swimming together in close contact with much rubbing and stroking, nuzzling and nibbling, much of which is directed to the genital region of the partner, much as a dog follows a bitch in heat and by his actions stimulates her into receiving him.

From the observations of former and present whalers it is generally accepted that the Sperm whale is polygamous, and that small schools of females are attended by a master bull, who is alleged to drive other adult bulls away. Nevertheless, although this species has a long jaw well armed with teeth and can open its mouth wide, scars that could have been made by bites from rival males have seldom been recorded. In the dolphins the scars are series of parallel lines made by some of the teeth of one side of the jaw. In most species they are not numerous, showing that encounters are not common, perhaps because the animals are not very aggressive towards each other when following females in heat. In a few species, however, extensive scarring is always found, and must point to a different kind of sexual behaviour. Such scarring with parallel lines is plentiful on the skin of Risso's dolphin, and is also common on the False Killer whale and on the Blackfish. The differences in behaviour causing these different effects have not yet been discovered.

The Beaked whales – the ziphiids – show a specially interesting pattern of scarring. As was pointed out in Chapter 2, the male ziphiids possess enlarged teeth or 'tusks', generally a pair either at the tip of the lower jaws, or set farther back towards the angle of the mouth. These 'battle teeth' as McCann aptly calls them are evidently used in fighting between males, for their unmistakable scars are conspicuous on the bodies of mature bulls. The scars are typical of each species; those on True's Beaked whale for instance, which has two battle teeth at the tip of the jaw, are always two parallel streaks, whereas those on species where the battle teeth are farther back in the jaw have scars in single streaks made by gouging with the tooth of one side at a time. The scars are a result of male rivalry, as is shown by their presence nearly always only on the skin of mature bulls but seldom of the females; McCann concludes that the bulls do all the fighting and 'in the mêlée may, perhaps, scar a female accidentally, which happened to be in the way'. The constant scarring, and the presence of battle teeth, show that the sexual behaviour of the ziphiids differs from that of the dolphins, and may well be correlated with the possession of schools of females by aggressive proprietor bulls. But ziphiids are seldom observed at close quarters in the wild and this conclusion can be no more than inference in the present state of our knowledge.

With the caution that there is no completely reliable way for finding the absolute ages of cetaceans – neither by counting corpora albicantia in the ovaries, laminations of the ear plug, ridges on the baleen plates, nor dentine layers in the teeth – we can agree that much useful information has nevertheless been accumulated about the breeding and growth of the animals. The general conclusion to be drawn from the numerous studies on the subject is that compared with most land mammals cetaceans are rather slow breeders, that the period between birth and reaching sexual maturity is comparatively long, as also is the total life span. On the other hand the rate of growth both before and after birth is high – in the large whales there is a very rapid rate of growth in the last two or three months of pregnancy and in the months of infancy after birth. The comparatively large size of the young at birth is attained by a high growth rate rather than by a long period of gestation – the larger the whale the faster its foetus grows.

Chapter 5

Swimming and diving

Anyone who has seen a school of porpoises or dolphins swimming round the bows of a ship will have been struck by the graceful and apparently effortless way in which they not only keep up with a swiftly moving ship but are able to shoot ahead or overtake it. They seem to flow through the water spontaneously, for it is impossible in the swirling water to see any movement of the flukes, their organs of propulsion. The explanation of this exhilarating sight, which has given pleasure to so many for so long, is still not fully understood though great progress has been made towards doing so.

The factors governing the swimming of dolphins differ from those pertaining to ships because the hull of a ship is only partly submerged whereas the body of a dolphin is completely below the surface except during brief moments of respiration. Both, however, are subject to the effects of skin friction with the water, which causes the smooth laminar flow at the front end to break up into a turbulent flow of eddies which greatly increases the drag that retards progress as the water passes aft. Thus the maximum economical speed in knots for a ship is 1.3 to 1.7 times the square root of the waterline length in feet; beyond this speed the turbulence and generation of waves is so great that increased power proportional to the cube of the extra speed is needed to drive it faster, and even this expedient is soon defeated by cavitation occurring in the water against which the propeller is thrusting. But for a completely submerged body such as a cetacean this formula does not apply.

Cetaceans can reach the high speeds they do through their ability to avoid the creation of turbulence in the water in contact with their bodies, so that the flow of water past them is laminar and drag is reduced to a minimum. In the first place the shape of their bodies is faired to a modified streamline form, but even with the best of streamlined bodies the drag varies with speed, and there is only one optimum speed for minimum drag. Cetaceans being living creatures are able to alter the surface characteristics of their bodies, so that they can accommodate them to different speeds with the greatest efficiency in eliminating drag. But before going further into these matters it is necessary to understand how cetaceans produce and use the power necessary to propel them.

Cetaceans, unlike the eared seals and the land mammals, use only their tails and not their limbs for locomotion – they do not possess functional hind limbs, and their fore-limbs in the shape of paddles or 'flippers' are suitable for balancing or turning but not for propulsion. The muscles of the cetacean body differ widely in their arrangement from those of other mammals, although comparative anatomists can trace homologies between the two. As would be expected, the main locomotory muscle masses are concentrated on the movement of the tail, the wide horizontally set flukes of which drive the animal through the water. The flukes oscillate up and down; it is the up-stroke that does the work, the return down-stroke bringing the flukes into position for the next driving up-stroke.

The action is similar to that used in sculling a boat with a single oar over the stern; although in that manoeuvre a man moves the oar from side to side, whereas a cetacean moves the flukes only up and down, it is the upward pressure of the oar, using the stern as a fulcrum, that drives the boat ahead. The side to side movement is quite irrelevant to the forward propulsion of the boat, and is needed only because if a straight upward power stroke is used it is impossible to follow it with a straight downward stroke, for the oar will come out of the sculling notch in the transom. The blade of the oar must therefore follow an oblique downwards path while at the same time it presses upwards against the water. Anyone who can scull a boat knows that it is the downward pressure of the hand at the tip of the loom that causes an upward pressure by the blade against the water on the other side of the fulcrum, and that there is no rotary motion. In a cetacean there is no need for a side to side movement because the upward power stroke can be followed by a passive downwards one with minimum resistance from the flukes. In both cases, however, the underlying principle is the same, an upward power stroke against the resistance of the water. Similarly a skin diver wearing 'frogman' feet uses the upward stroke for propulsion.

The locomotory muscles of a cetacean are fixed at their inboard ends to the vertebral column, and their outboard ends terminate in tendons

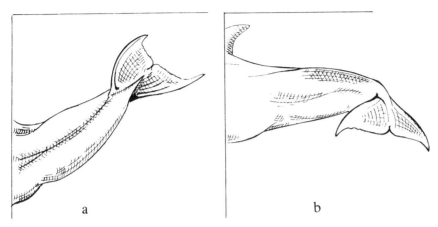

Figure 14. The tail of a Bottle-nosed dolphin in successive swimming positions showing (a) the passive down stroke, and (b) the power up stroke.

fixed to the tail vertebrae, the farthest of which lie in the base of the flukes. The muscles are arranged in two main masses, the epaxial mass which lies along the upper side of the backbone and serves to raise the tail, and the hypaxial mass below, which draws the tail downwards. Whalers know these masses as the 'back muscle' and the 'under fillet'. In large whales these masses weigh many tons, and to those whalers such as the Japanese who value whale meat more highly than whale oil they provide magnificent cuts – huge chunks of red meat without any bone or any inedible trimmings.

In the heyday of Antarctic whaling the cook of a whale catcher would get a lump 4 or 5 feet long and about a foot square in section, and hang it in the rigging for a week. By that time the outside turned to a black solid crust, which only needed to be sliced off to leave a mass of red, well hung beef ready for cutting into juicy steaks. The European hunters of the 'small whale fishery' who catch mainly Minke whales with some Killers, Bottle-nosed whales, Blackfish and others, send this excellent meat which should be feeding hungry men to the pet food manu-facturers, whence it appears recycled as the dog-droppings that foul city pavements, or goes to nourish pampered cats that ought to be earning their livings by catching mice and rats. When enterprising Asiatic eating-houses in London are caught serving pet food in curry they are prosecuted.

The epaxial muscle mass is much larger than the hypaxial, for it is the one that raises the tail and does the work in driving the animal forward. At each stroke the flukes turn out of the horizontal position, downwards on the up-stroke and upwards on the down-stroke, as would be expected from the pressure of the water when the tail rises and falls. Part of the epaxial mass is separated from the main mass by a tough septum of connective tissue so that it can be distinguished as the fluke-elevator muscle between the epaxial and hypaxial masses. Its tendons are fixed to the last few vertebrae that lie in the base of the flukes ([148]. The action of this muscle is to brace the flukes against the downward pressure of water on the up-stroke so that the correct angle of

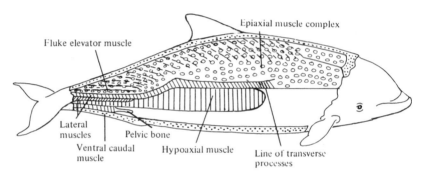

Figure 15. Diagram showing the positions and relative sizes of the swimming muscle-masses of a dolphin.

attack is maintained to push the body ahead. There is no corresponding depressor muscle to brace the flukes on the downward stroke, and consequently the flukes passively pivot upwards in response to the water pressure as the tail is brought down into position for beginning the next upward power stroke. It has been pointed out, however, that although the hypaxial muscle mass is smaller than the epaxial, the muscles of the belly wall are also fixed to the tail and together with the hypaxial muscles approach in combined weight that of the epaxial. This could mean that the downward stroke also contributes power to produce forward movement. At slow speeds and during acceleration the amplitude of the tail and fluke oscillations is large, but at higher speeds the amplitude of the movement needed to maintain constant speed is comparatively very small.

Several cetologists have calculated the energy needed for swimming in various cetaceans. Assuming that the conversion efficiency of cetacean muscle is 25 per cent, a dolphin weighing 100 kilograms consumes about 0.1 kilocalories per kilogram of weight for every kilometre travelled at 10 knots, if it swims through the water with a laminar flow and no turbulence [180]. A Fin whale 18.1 metres long and weighing about 34 tons, swimming under similar conditions at 10 knots, uses 0.022 kilocalories per kilogram for each kilometre travelled, and a Blue whale weighing 100 tons similarly expends only 0.014 kilocalories per kilogram for each kilometre. The total energy expended by the 18.1 metre Fin whale in going one kilometre is about 748 kilocalories in all [98]. Thus the bigger the whale the less energy is needed to drive it through the water as it swims, a matter of importance to them on their long migrations, which we shall consider further in Chapter 6.

The calculations of energy consumed in locomotion are based on the assumption that the flow of water over the surface of the body is laminar so that there are no turbulent eddies to create drag. In spite of all the improvement in hull design it has not been possible to build a surface ship that does not create turbulence, as a glance over the side of a moving vessel at once shows, but the body of a cetacean, being alive and flexible, can to a large extent avoid producing turbulence, so that the water-flow over the surface is nearly laminar.

The problem of how cetaceans reach the high speeds they do is particularly interesting in the smaller species, the porpoises and dolphins. The large whales normally cruise at no great speed, about 5 knots may be the average, but when they are frightened as by a pursuing whaler they can sprint at 20 knots or more. Dolphins less than a tenth of their length and less than a thousandth of their weight can reach speeds approaching if not equalling theirs. The power output of mammalian muscle, as judged by the maximum sustainable effort in man, is not much more than 0.1 horsepower per pound. The weight of the muscles of a large whale is enough to produce the power needed for attaining their known speeds, but in the dolphins it is much less when compared with the power needed to drag a rigid body of the same shape at the same speed through the water. The power output of a dolphin's muscle if it were a rigid body would have to be six or seven times as great as that of a large whale, because as size increases the amount of muscle increases with the cube of the body length, whereas the surface resistance increases only with the square [73].

The high speeds of dolphins, which appear to be greater than the power output their muscles can produce, has been termed a paradox

[150]; it is not, however, a paradox but a fallacy. The fallacy arises from assuming that the power needed to move a dolphin is the same as that needed to move a rigid body of similar shape and size. But the body of a dolphin is living and flexible, not dead and rigid, so that when the properties of the living dolphin's body are examined it is found that there is not a paradox, as indeed there cannot be if the phenomena are correctly observed.

The outer skin of cetaceans is very thin and closely applied to the thick underlying blubber which forms a case wrapping the whole body. The blubber, however, is not tightly fixed to the underlying muscular tissues, so that there can be considerable movement of the one over the other. The loose attachment of the blubber is seen when a whale carcase is flensed – lengthways cuts in the blubber allow it to be stripped off by winch like peeling a banana. The blubber hardens after death, but in life it is softer and easily distorted by external pressure.

The American cetologist Frank Essapian was the first to record the occurrence of skin folds on the surface of dolphins when accelerating or moving at high speed [54]. The folds take the shape of transverse corrugations mainly along the sides and under surface of the body, and appear for a few seconds only during strong acceleration and deceleration, and for longer periods during swimming at high speed. They are not produced by direct muscular action but by different pressures of the water at different parts of the body surface, the ridges corresponding to spots of low pressure and the dimples to ones of high pressure. Any tendency to produce turbulence at the body surface is thus automatically counteracted by the response of the blubber to variations in pressure. When the dolphin is swimming fast the folds remain stationary and do not pass along the body in a wave-like manner, showing that the spots of high and low pressure remain constant in position. Essapian noticed that at extremely high speeds the folds tend to slope backwards, an important point discussed below.

The speed-induced skin folds differ from the folds or wrinkles that occur mainly at the throat and side of the head when the animal turns its head, as it can do to a surprising extent in life, and would not be expected from examination of its dead body. The folds and wrinkles are much like those produced in the neck of an obese seal, or in a person with a double chin.

Another line of research by the British cetologist Dr P. E. Purves has gone far to elucidate the way in which dolphins obtain a laminar flow of water over the body surface at high speeds [159]. Although it is almost impossible to measure the water flow round the body of a submerged free-swimming dolphin, he was able to reach some interesting conclusions from observations on dolphins at the surface, with bodies partly exposed during breathing, and especially from some remarkable photographs, taken by Dr F. C. Fraser, of the Common dolphin in the North Sea. A speeding dolphin in smooth water, with about half the back and the dorsal fin exposed, shows a bow wave spreading out in an extremely thin sheet of water on each side of the body 'like the wings of a butterfly'; behind, there is a powerful upflow of turbulent water above the tail. Purves deduced that when the animal is submerged the two sheets of water of the bow wave would meet and close over the back, and the upwash at the back would be a powerful acceleration of water over the upper surface of the flukes. Furthermore, the shearing stresses at the surface of the skin due to the flow of water would be greatest in a fore-and-aft direction; an examination of the detailed structure of the skin supported this conclusion.

Although the surface of the skin in cetaceans appears smooth there is, nevertheless, a very well-developed system of dermal ridges beneath it, like the ridges on the palms of human hands and fingers that produce fingerprints. They serve in us to prevent the surface being stripped off by shearing stresses, as when we grasp the handle of a tool. Purves was able, by the patient use of abrasives, to remove the outer layer of the skin, the stratum corneum, of a Common dolphin and a Common porpoise, and expose the dermal ridges for examination. In the dolphin most of the ridges on the sides run backwards inclined at an angle of about 30° with the long axis of the body, a narrower band runs straight fore-and-aft below them, and on the under surface of the body they are very poorly developed. In the Common porpoise, the slower swimmer, the general pattern is similar but the ridges on the back and sides are more fore-and-aft as far back as the tail, where they too have an upward slope. The direction of the ridges corresponds with that of the water-flow over the surface – and here the probable meaning of the speed-induced skin folds that slope backwards, seen at high speeds by Essapian, becomes apparent.

Purves concluded that as the upward stroke of the tail and flukes is the power stroke, it will tend to deflect the head end of the animal downwards, and that the under surface of the head and belly where the skin ridges are scanty is the area subjected to the consequent pressure –

the lift produced is equivalent to planing. The front-to-back flow of water round the hydrofoil of the flukes during their powerful upthrust accelerates the water-flow over their upper surface, and produces an obliquely upward flow over the flanks, as is shown by the direction of the skin ridges. It also helps to return the flukes to a horizontal position on their passive down-stroke. Finally, Purves calculated as a result of his work that the oblique flow of water over the body of a 6 foot dolphin at 15 to 20 knots would be 90 per cent laminar. There is thus no paradox about dolphin swimming – the structure and movements of the body ensure laminar flow, unlike the inert rigid mass of an equivalent artificial model or even the dead body of a dolphin if towed at similar speeds.

There is yet another way by which the turbulent flow of water over a moving submerged body can be reduced. It is now known that the addition to the water of extremely small amounts of high polymers with long-chain molecules of high molecular weight dramatically reduces the drag. The addition of only twenty parts per million of an unbranched polyethylene oxide with an approximate molecular weight of one million reduces the drag on immersed moving bodies and ship hulls by over 50 per cent. The coiled macro-molecules are deformed in the turbulent vortices and cause an extremely high dissipation of energy in them, thus increasing the stability of the surface layer and reducing the growth of vortices. A similar though smaller effect is produced by suspensions of minute solid particles [153]. This effect appears to be made use of by fishes, which secrete a layer of mucus that covers their skin – the most effective secretors are predators that depend upon quick acceleration and high speeds to catch their prey. In shoaling fish the mucus streaming from the leaders reduces the effort needed by the followers in keeping station. In an experiment the addition of about forty-two parts per million of a high polymer ethylene oxide to the water nearly doubled the swimming speed of the fish [4].

The outer layer of the skin, the epidermis, of delphinids is unlike that of most mammals in that it has a firm, smooth outer layer consisting of cells that are not heavily keratinized to form a covering of non-living horny matter. Using the scanning electron microscope Harrison and Thurley [81] have shown that the layer is cellular with nuclei almost to the surface, where the cells become condensed and flattened. Droplets of oily material are produced in some of the cells and are discharged from their surfaces. In addition, although the skin surface is smooth the outlines of its cells can be made out by the scanning electron microscope, which shows that they are desquamated in large numbers – they are continually being detached and released into the water in contact with the body. It is possible that the secreted droplets and the detached cells entering the water-flow over the surface help to maintain a laminar flow and reduce turbulence and drag by dissipating the energy of the impeding vortices. On the other hand the Russian cetologist Sokolov and his collaborators [195], who made some experiments with cells scraped from cetacean skin, concluded that the cells shed from the epidermis had a negligible effect on hydrodynamic drag.

Finally, to return to the first paragraph of this chapter, the wave-riding of dolphins in the bow wave of a ship. This is not part of their repertoire in the normal course of their lives, for they appear to do it mainly in the bow waves of ships and have only rarely been seen to ride the forward slope of wind waves. The riding of wind waves, however, may be commoner than is thought, owing to lack of observers and the difficulty of seeing the happening at anything but very close quarters. In the bow wave dolphins keep station under the fore-foot of the ship, and when they sometimes turn on their sides it can be seen that they are not swimming by oscillations of the tail flukes, which remain perfectly motionless. Many naturalists had speculated about how they do this until Dr Scholander of the Scripps Institution of Oceanography solved the riddle by experiment [182]. He made an artificial fluke that could be lowered into the bow wave and was connected with a dynamometer that showed the drag or push derived from the wave. When the trailing edge of the fluke was tipped upwards at an angle of 28° it gave a considerable forward thrust when the ship was going at 8 knots, but at other angles it developed a heavy drag. A dolphin is thus pushed forward in the bow wave by putting its tail fluke at the correct angle into the upwelling water while planing its body forward horizontally in the ship's course. Scholander concluded that if a dolphin steers itself horizontally and 'leans' the tail flukes against the upwelling water of the bow wave, it cannot help but be pushed along with the ship. As the water is thrust not only upward and forward but also outward, it can ride heeled over on its side as well as on an even keel. 'Moreover, as this mode of propulsion does not require that his lungs be empty [to have negative buoyancy] he need not take his ride in silence but may whistle to his fellow freeloaders as much as he deems fit.'

The German aeronautical engineer H. Hertel [83] has studied the mathematics of this problem, and points out that dolphins can ride the waves only because they are such swift swimmers that they can reach the critical speed necessary to position themselves in the forward slope of a rapidly advancing wave. Furthermore, the forces acting on the tail fluke are only a partial contribution to riding the wave; the dolphin uses not only the propulsion and pressure field of the wave but positions itself so as to use the velocity field in the forward slope of the wave to stay in equilibrium with it and to be carried ahead.

Another interesting point emerges from this work. Dolphins have to surface frequently to breathe, and when travelling at high speeds they have to jump clear of the water in order to do so efficiently. If they merely expose part of the body to bring the blow-hole above the surface, as they do at lower speeds, the turbulence created is so great that the resulting drag would slow them below the critical speed for wave-riding – they would in effect be subject to the same limitations as those applying to a surface ship. The jump has a flat trajectory so that it is quite unlike the breaching of the large whales, or the jumps of the Spinner dolphin, but resembles the porpoising of penguins which similarly jump out in a low trajectory to breathe when swimming fast. It is a manoeuvre to avoid meeting maximum resistance through partial emergence. Thus what appears to us to be a joyful and sportive exhibition of high spirits is in reality nothing of the sort; contrary to the anthropomorphic interpretation it is merely a matter-of-fact necessity.

All the researches we have been considering give some explanation of the hydrodynamics of the swimming of cetaceans and the way the smaller ones can move at such surprising speeds. There now remains for examination the way in which they are able to withstand the stresses of deep and prolonged diving. Here again we find that there has been much muddled thinking leading to false conclusions, through approaching the problems from the wrong direction and considering them the same as those that confront a human diver.

When a man uses diving equipment, either the old style 'copper knob' or 'hard hat' scaphander, or the modern skin-diving 'scuba' self-contained underwater breathing apparatus, to reach depths greater than his breath-holding endurance will allow, or to remain below for a great length of time, he breathes compressed air. This is necessary because the air pressure within his lungs must equal or slightly exceed the pressure of the water around him; were it less his chest would be crushed by the external pressure. Furthermore, in order to breathe out, the internal pressure must exceed the external so that the air he has used may escape through the exhaust valve in the familiar stream of bubbles. The underwater pressure increases by one atmosphere or 15 pounds per square inch for approximately every 10 metres of depth; thus at 30 metres, or 100 feet, the pressure is four atmospheres or approximately 60 pounds per square inch.

A diver has no difficulty in breathing compressed air or in coping with the increasing pressure as he descends, but when he starts returning to the surface from more than a moderate depth he runs into trouble. While he is under compression the nitrogen of the air dissolves in the fluids and tissues of his body to their full capacity at the incident pressure, and their full capacity at that pressure is much greater than it is at the surface under a pressure of only one atmosphere. Consequently when he ascends and the pressure diminishes the dissolved nitrogen comes out of solution in the form of bubbles of the gas. The process is exactly similar to the bubbling of soda water when the pressure is released by opening the bottle.

The bubbles of nitrogen may appear in any part of the body – in the blood vessels to cause a gas embolism, in the joints to cause a painful condition called 'the bends', in the nervous system to cause paralysis or 'diver's palsy', and elsewhere. These conditions are always painful, and can be fatal; to avoid them the diver must ascend very slowly so that the excess nitrogen is voided gradually in the breath without forming bubbles in the tissues, or he may enter a decompression chamber and be hauled up to the surface in it for the pressure to be gradually lowered. There are other difficulties at great depths such as the toxic effect of breathing air with a high partial pressure of oxygen, but the nitrogen bubbles that appear on quick decompression are the most important for present purposes.

For long zoologists were puzzled about the way whales could avoid getting the bends or other symptoms of caisson disease or compressed air sickness, as it is called, and various theories were propounded to explain how the excess nitrogen dissolved in their tissues could be removed. They need not have exercised their ingenuity, for the solution of their problem is absurdly simple: cetaceans do not get the bends because they do not breathe compressed air. When a cetacean dives it cannot take with it more air than will fill its lungs, and of this rather less than four-fifths is nitrogen. The capacity of the lungs in proportion to the body weight is comparatively small in cetaceans; it varies from 1 to about 3 per cent whereas in man it is over 7 per cent, so that the amount of nitrogen that could dissolve in the body fluids and tissues from one filling of the lungs is also comparatively small.

Cetaceans, however, are free from the danger of even this small quantity's becoming dissolved in their blood and tissues, through the effect of increased pressure on their bodies. The lining of the alveoli, the ultimate air passages, in the lungs of all mammals is exceedingly thin, so that the blood in the capillary vessels is as nearly as possible in contact with the air from which the haemoglobin takes up oxygen and into which it gives up carbon dioxide. It is through this tenuous interface between air and blood that the unwanted nitrogen invades the blood in a man breathing compressed air. When a cetacean is at the surface and breathes air at atmospheric pressure the alveoli of its lungs are similar, but when it dives they undergo an important change.

The body of a cetacean, or of any other mammal, is not compressible because it consists largely of water which permeates all the tissues; and water is practically incompressible. When a cetacean dives, therefore, the external pressure is transmitted equally to all parts of the body, which suffers no distortion; the only contents that can be compressed are the undissolved gases, mainly the air in the lungs. When this air is compressed by the external pressure of the water during a dive it decreases in volume, as also necessarily do the lungs containing it. In cetaceans the chest is comparatively flexible and the diaphragm or midriff is set very obliquely and less directly transversely across the body than in land mammals; thus the pressure of the abdominal viscera pushing against it on one side makes the lungs on the other side collapse and drives the air in them into the windpipe and its branches and into the extensive nasal passages. As a result the alveoli of the lungs are emptied of air and the membrane lining them is greatly thickened, so that the exchange of gases between the blood and air is reduced and with it the invasion rate of nitrogen.

The deeper the cetacean dives the more its lungs collapse and the more the invasion rate of nitrogen decreases. At a pressure of 11 atmospheres, or 165 pounds per square inch, at a depth of 100 metres, or 328 feet, the lungs are completely collapsed and contain no air, all of which is pushed into the more rigid parts of the respiratory tract where gas exchange is not facilitated as it is in the alveoli of the lungs. Once a cetacean has reached this depth, where the invasion rate of nitrogen is reduced to zero, it can with perfect safety go to any depth beyond. In contrast, the lungs of a human diver, filled with air at a slightly greater pressure than that of the surrounding water, do not collapse, so gas exchange continues, and the invasion rate of nitrogen goes on increasing the deeper he goes. As a cetacean returns to the surface the lungs gradually expand again so that when it blows and fills its lungs with fresh air the alveoli have returned to their expanded condition by which a high rate of gas exchange is attained.

The ventilation of the lungs in a cetacean is highly efficient, in spite of the rapidity of the air intake in the intervals between the holding of the breath. The large whales in particular, on returning to the surface from a long dive, draw a number of breaths before going down again – the old whalers hunting Sperm whales called this 'having his spoutings out'. A rorqual usually starts blowing – exhaling – the moment its blow-hole breaks surface, often a trifle before, and then very quickly inspires. The widely opened blow-holes appear surprisingly large at this moment; the longitudinal slits that they form when closed expand to large pipes and are raised above the surface of the head so that a sort of cutwater prevents water splashing into them. The 'blow' is a cloud of steam produced by the condensation of the moisture in the warm breath on coming into contact with cool air. This is obvious enough in cold latitudes, but it may be surprising that the blow is equally visible in the tropics. The exhaled air, however, is saturated with moisture so that even when blown out into warm surrounding air there is some condensation, just as the exhaust from a steam engine is visible in the hottest climate. Another thing also probably reduces the temperature of the exhaled air; when a gas expands under reduced pressure its temperature is lowered, a process known as adiabatic cooling, which is particularly obvious when a gas under pressure is released through a small nozzle. When a whale blows the breath is exhaled with some force from the nostrils – necessarily in view of the short time available for both exhaling and inhaling – and this alone must cause considerable adiabatic cooling with consequent condensation.

A further matter may contribute to the visibility of the blow. The extensive air-sinuses connected with the nasal passages in the head of cetaceans, which are described in Chapter 7, are filled with a foam consisting of an emulsion of oil and mucus [66]. It has been suggested that some of this foam gets mixed with the exhaled air stream and thus helps to make the blow visible. Furthermore, the oil in the emulsion is a good solute for nitrogen, and it is possible that it serves to remove any excess nitrogen that may have invaded the tissues. Be this as it may, it is fairly certain that some of the foam does get into the blow for, as every whaler knows, the fishy-smelling breath of a whale is anything but sweet.

Another question arises about the diving of cetaceans: how do they manage to keep their metabolism going, and especially their muscles working, for comparatively long periods on only one breath – a single filling of the lungs? Do they store oxygen in their tissues, or are they able to function anaerobically on a short supply of oxygen? The volume of blood relative to body weight is greater than in most land mammals but is not nearly as large as it is in seals; and the haemoglobin has no greater affinity for oxygen than in other mammals. As far as is known the heart rate does not slow down as drastically as it does in seals when they dive. Unlike seals, too, they do not during a dive accumulate a large pool of venous deoxygenated blood in the abdominal vessels so that the limited amount of oxygenated blood is reserved for the brain, which is quickly and permanently damaged by lack of oxygen.

In the first place the muscles of cetaceans contain an unusually large amount of myohaemoglobin – hence the dark colour of whale meat – which combines with oxygen and forms in effect an oxygen store. Myohaemoglobin is found in all mammalian muscle, but in cetaceans the quantity is up to ten times as much as is usual in land mammals, so that on diving the muscles take with them a large supply of oxygen. When this store of oxygen is used up the muscles are indeed able to go on working anaerobically for longer than the muscles of land mammals, because cetaceans are slower in reacting to the amounts of lactic acid and carbon dioxide that pass into the circulation as a result of muscular activity. These substances stimulate the respiratory centres of the brain, leading to compulsive respiration.

There is, in addition, a peculiar arrangement of blood vessels in many mammals which is particularly elaborated in the cetaceans. This takes the form of networks of blood vessels communicating with each other massed together to form complicated tangles of some size, the retia mirabilia. They take various forms such as arteries communicating with veins, arteries communicating with each other, and other varieties. They usually occur in relation with the joints of the limbs, or at the base of the brain in the skull, but in cetaceans they are much more widespread, and are particularly conspicuous as arterial retia on the inside wall of the chest and between the ribs. The function of the retia is not properly understood, though as they are so conspicuous and massive in the cetaceans they are presumably in some way connected with diving. It has been suggested that they act as reservoirs of oxygenated blood, but the total amount of blood that they can hold is not enough to be useful in a long dive. It may be that they regulate the flow of blood through the brain, those in the skull ensuring that it is supplied with oxygen when the other parts of the body are deprived, or it could be that they are concerned with counteracting any adverse effects of increased pressure at depth, though as the pressure is bound to be equal throughout the tissues of the body it is not clear what adverse effects may be produced. Until further researches have been made it seems unwise to be dogmatic about the part the retia play in the metabolism of cetaceans.

The superficial blood vessels of cetaceans, the small vessels that pass through the blubber, break up into minute capillary vessels beneath the surface of the skin. The arteries carrying the blood outwards run close to the veins carrying it back, so that the veins filled with cooler blood take up some of the heat from that in the arteries and take it back into the body; this heat-exchanger arrangement helps to conserve the heat of the body. It has been stated that the temperature of a large Blue whale would rise several degrees during a ten-minute burst of speed at 20 knots, and that the superficial blood vessels would then dilate to bring an increased amount of blood to the surface so that excess heat would be lost to the water. On the other hand figures of doubtful reliability have been produced that claim to show the insulating value of the blubber is not enough to prevent excessive heat loss, and that a whale must keep moving to keep warm [149]. Those who consider such a state of affairs to be highly improbable in view of the close adaptation to their environment that the cetaceans have achieved during several million years of evolution, point out that on the contrary the thickness of the blubber in different species points to the likelihood that the thermal insulation of the blubber is adequate. It is not so thick in the active rorquals as it is in the more slowly moving Right whales, conditions that would be expected if the amount of insulation needed is related to the production of excess heat by muscular activity.

A further point of interest is found in the skin of cetaceans. Although the body is hairless, all cetaceans have a small number of 'whiskers', at least during foetal development. These correspond with the vibrissae of other mammals, the long hairs on the snout but also found elsewhere, that are connected with nerve endings and serve as sensitive organs of touch, and possibly have other functions as well. The hairs on the head and jaws of cetaceans are similarly supplied with nerve endings, and even when the hairs themselves are lost or not fully grown during foetal life, the follicles and their nerve end organs remain. The English anatomists Palmer and Weddel have closely investigated the nerve endings of the skin in the Bottle-nosed dolphin, *Tursiops truncatus*, and infer that the conditions they found are similar in other cetaceans [146]. They find that in the snout of the adult the hair follicles of the newly born calf have become modified into richly innervated pits, which are specialized sense organs connected with the rapid perception of the smallest changes in water pressure in their immediate neighbourhood. The pits are adapted to signal pressure changes by the magnitude and direction of the displacement they undergo as a result of these changes. The researchers suggest that the end organs detect the relative speed at which the animal is swimming – they are in effect speedometers. They also found that cetacean skin contains many other types of complex, circumscribed, often encapsulated nerve endings in the skin of other parts, in addition to those on the snout. These also are concerned with the perception of pressure differences, and probably serve in adapting the surface of the body during swimming to achieve a laminar flow of the water passing over it. Such surface changes would be reflex and automatic, so that they would not be consciously produced by the animal. The innervation of the skin is richer, more elaborate and more specialized than the skin from comparable regions in man.

Cetaceans dive, among other reasons, to seek and capture their food; how therefore do they prevent the water entering the lungs when they open the mouth underwater? In all mammals the upper end of the windpipe is formed from several cartilaginous plates that together make the larynx, a box of irregular shape opening in front of the pharynx at the back of the tongue; when the animal swallows, the food or liquid has to pass over the opening to reach the oesophagus leading to the stomach. The act of swallowing raises or pushes forward the larynx, and at the same time the front cartilage, the epiglottis, folds down backwards to close the opening and prevent the food from 'going down the wrong way'. But in whales at depth, with the lungs collapsed, the air pushed out of the lungs is compressed in the windpipe and in the nasal passages, with their ramifications between the blow-hole and the inner opening of the nasal passages in the roof of the pharynx. In all cetaceans, and particularly in the odontocetes, which are known habitually to go to

...ly engraving of the Sperm whale stranded at Scheveningen in the Netherlands in 1598. Numerous ...were made from this picture for over two hundred years.

...graphic plate illustrating Flower's paper on the Sperm whale in the *Transactions of the Zoological* ...*of London* (1868). Figure 1 shows the skull of a Sperm whale cut in half; note the 'Neptune's chair' ...1 by the crests of the maxilla and supraoccipital, the small size of the brain-case, and the hollow ...aw. Figure 2 is a similar preparation of the skull of a very young Sperm whale, and Figure 3 a ...˙ one of the skull of a nearly adult Bottlenosed whale.

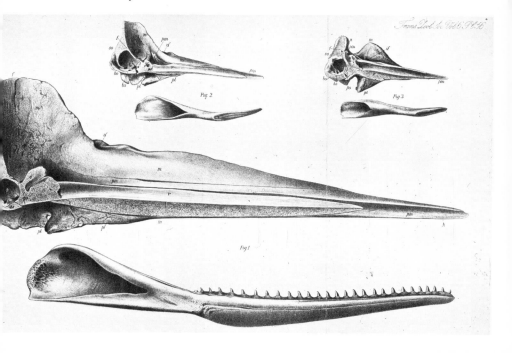

A dead Fin whale at Durban whaling station. The animal lies on its right side, and shows the groov the underside, the asymmetrical colour pattern and some of the baleen on the roof of the mouth.

The roof of the mouth of a Sei whale that is being cut up at a whaling station. The animal lies on its b and the lower jaws have been removed; tip of the snout to the right, throat to the left. The narrow pal lies between the two sides of baleen, the fibrous inner fringes of which make the filter bed.

e whale blowing. The whale is poking its head out of the water because it is hemmed in by the
ice floes. The blowhole is just above the water on the right, and the throat grooves can be seen
he lower jaw to the left.

of the snout of a Black Right whale coming up to blow; the blowholes have not yet broken the
The bonnet, the mass of horny skin covered with barnacles, lies close to the tip of the upper jaw.
scattered hairs on the upper jaw in front of it.

Part of a school of Killer whales putting their heads out of the water to look at the photographer who standing on an ice floe in the antarctic.

A tame White whale in an aquarium. The whale can control and alter the shape of the melon, which i this picture is maximally expanded.

Part of a school of White whales photographed from the air in the Canadian arctic. The whales are swimming in shallow water, and many of them are accompanied by calves.

A Gray whale pushing the end of its snout out of the water, the lower jaw is to the left. Note the growth of barnacles on the left side of the upper jaw.

ck Right whale breaching. Note the broad flipper and the bonnet at the end of the upper jaw.

mpback breaching backwards. Note the long narrow flipper with knobs on the leading edge, and
s on the upper surface of the rostrum. The eye is seen at the corner of the mouth, and the coarse
t grooves end at a lump under the chin called the 'cutwater' by whalers.

Pacific Bottlenose dolphins leaping clear of the water to blow at speed.

Two captive Indian Freshwater dolphins (*Platanista*). Note the long narrow beak, the broad flippers a the rudimentary eye. The lower animal is characteristically swimming on its side near the bottom; it al shows the unusual longitudinal shape of the blowhole which is tightly shut.

mmerson's dolphin, the most strikingly marked species of the genus *Cephalorhynchus*; small dolphins
h inconspicuous beaks, from the subantarctic and southern South American waters.

underwater photograph of Pilot whales or Blackfish. The prominent melon gives them their whalers'
ne of 'Pothead'.

A Bottlenose dolphin (in an aquarium) giving birth. Note that the young is born tail first.

ough the young was born dead the mother tried to lift it to the surface in the usual way to help it take rst breath. The broken umbilical cord has not yet been expelled with the afterbirth.

ky dolphins, *Lagenorhynchus obscurus,* porpoising at speed. The blowholes are wide open as the hins inhale.

The underside of the chin of a dead Humpback, showing the coarse throat grooves with barnacles on t
ridges between, ending at the cutwater at the front on the left.

Erotic 'play' of a captive male Bottlenose dolphin.

The mass stranding of a school of Pilot whales on the coast of Orkney.

greater depths than the mysticetes, there is a special arrangement that ensures the continuity of the nasal passages and the windpipe so that air is not lost and water cannot enter the lungs when the mouth is opened. The upper part of the larynx is elongated and sticks upwards in the form of a tube with the free end inserted into the posterior nares, the inner end of the nasal passages, where it is firmly grasped to prevent its slipping out.

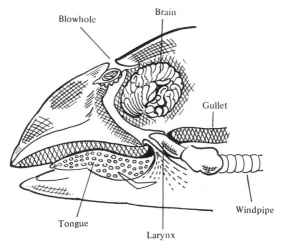

Figure 16. The head of a porpoise bisected down the centre, showing the larynx projecting into the posterior nares leading to the blow-hole, and the air sacs inside the blow-hole.

The deeper a whale dives the longer must be its ability to refrain from respiration, and the greater the 'oxygen debt' that it incurs – not a very good term for its owes oxygen to no one, it is rather a credit that it draws upon on its return to the surface. The rorquals do not need to go to great depths in search of their food, which is often quite near the surface. They normally dive for periods of five to fifteen minutes between coming up for air, though they can endure for half an hour or more if they are frightened. The Sperm whale, which can go to great depths, dives for thirty minutes to an hour or more and so incurs a greater oxygen debt. The number of breaths taken at each return to the surface to replenish air supplies is thus four or five times during about five minutes in the rorquals, but twenty to forty or even sixty over ten minutes or more in the Sperm whale. Rorquals probably do not habitu-

115

ally go to depths greater than some 300 feet, but Sperm whales find their prey at great depths – the bodies of Sperm whales have several times been found entangled in broken submarine cables recovered from depths of no less than 3,000 feet in the Pacific Ocean.

It has also been suggested that Sperm whales can alter their buoyancy as an aid to deep diving. M. R. Clarke found [37] that small changes of temperature at about 30 °C. produce large changes in the density of spermaceti, and therefore postulates that during the blowing at the surface the spermaceti is cooled by the inspired air so that buoyancy is decreased for a deep dive, and that when the whale is ready to return to the surface the blood supply to the spermaceti is increased so that it is warmed by the blood heat, its density decreases, and the whale has increased buoyancy to help it up. During a dive variations in the blood supply adjust the buoyancy so that it is neutral, and the whale needs to expend no energy in keeping at the depth required. This ingenious theory has not yet been proved experimentally, and is not accepted by many cetologists.

Even in the rorquals the oxygen debt is considerable because it has been calculated that the total of the oxygen stores taken down by a Fin whale 70 feet long is about 3,350 litres, an amount that suffices for a dive of sixteen minutes at a swimming speed of 5 knots. But, as such a whale can dive for at least twice this length of time and attain much greater speeds, it is apparent that a large oxygen debt must be built up. A harpooned Fin whale has been known to take a depth gauge down to over 1,000 feet [181], though it is doubtful whether it would normally go so deep. The smaller dolphins have not been accurately recorded as going deeper than about 500 feet.

In spite of all the information that research has been able to piece together, our knowledge of what cetaceans do in their underwater environment, and how they do it, is manifestly imperfect. We may hope that when submarine laboratories and other apparatus have been improved in quality and quantity cetologists may be able to follow the subjects of their study into the depths, and see by direct observation how they live their lives in those almost inaccessible regions.

Chapter 6

Migration

There are no territorial boundaries in the sea – at least as far as whales are concerned – and there is no physical barrier to prevent any marine cetacean from going to any part of the world, yet each species has its own range of distribution from which it does not normally stray. The areas covered are often large, but that does not imply that any individual may travel to any part of it, for species generally form populations which are restricted to their own particular part of the range.

Since the earliest days of whaling the whalers have known that their quarry appears at different places at different times of the year – that whales migrate seasonally from one part of their range to another. We now know that the migrations are closely correlated with the two fundamental necessities of cetaceans, as indeed of all animals, those of feeding and breeding.

The seasonal migrations of the mysticetes are the best known for several reasons: first because of the commercial importance of the species to the whaling industry, and further because the food of the mysticetes swarms in incredible quantities at certain seasons in the colder waters of the oceans, and because for most species a warmer and less rigorous environment is needed for the survival of the young at birth and during early childhood. Consequently they make annual journeys of great length between their feeding and their breeding grounds. The migrations of the rorquals inhabiting the southern hemisphere have been studied not only by direct observation but also by marking experiments. In the days of hand-harpoon whaling the whalers sometimes found old harpoons sticking in their victims which had escaped from and survived a previous attack. As harpoons were marked with the owner's name it was sometimes possible to trace when and where the whale had first been struck. Whale marking experiments were the direct descendants of these early unintended markings.

The whale mark was developed by the *Discovery* Expedition and first used in the Antarctic in 1926. The mark was like an enormous drawing pin or thumb tack, 3 inches in diameter, with three barbs at the point of the pin. It was shot with a blank cartridge from a twelve-bore gun, and to make firing possible a wooden stick pushed into a socket in the centre of the disc slipped into the barrel of the gun and made a gas-tight seal with wads attached to it. The stick also acted as a tail to make the dart fly straight, and broke off when the mark hit the whale and stuck in the blubber with the disc flush with the surface.

The marks were fired from a whale catcher or similar craft suitable for pursuing and approaching whales, and the species, approximate size, place, date and other details were noted. A reward was offered for the return of any marks subsequently recovered when a marked whale was captured commercially and reported with information about the date and place of its capture. A small number of whales was easily marked when the darts were first used, but it was only beginner's luck, for later experience found many defects in the design of the darts – the most serious was that the marks soon came out of the blubber, much as a thorn broken off in the skin of an animal works its way out with the passage of time. This led to the design of an improved pattern which consists of a stainless steel tube 10½ inches long, stamped with instructions and offering a reward for its return. This sort does not need a stick to make it fly straight, and does not have a disc to show on the surface. It penetrates deep into the blubber and often goes right through it into the underlying back muscles. It appears to do no harm to the whale, which is probably unaware that it has been pricked. These marks stay in the whale better than the first design, but have the disadvantage that they cannot be found until the whale is flensed. When they were embedded in the muscle they were often not found until the boilers were emptied after the oil had been boiled out of the meat, and then it was impossible to know from which whale it came and so no details of the whale's length or condition could be supplied. Cetologists of several nationalities have tried darts of different design, none of them with much success.

Whale marking is a most expensive undertaking, for it needs a special ship and crew, and if carried out on the whaling grounds can be frustrated by the capture of the marked whales soon after they are marked. The rate of recovery is also disappointing; only 370 of over 5,000 marks put into whales by the *Discovery* investigations up to 1939 were returned. Other marking campaigns have been carried out in New

Zealand and Peruvian waters, and by the Japanese in the north Pacific. The low rate of returns, and the great expense of marking campaigns, however, has not encouraged those concerned to continue large-scale experiments.

Although the results of whale marking have been on the whole disappointing they have, nevertheless, provided some valuable information, especially in confirming that whales regularly migrate between the tropics and subtropics and the Antarctic, and in establishing some knowledge of the routes along which they travel. The returns from temperate and warm waters, however, come from land-based whaling stations, for the pelagic floating factories do not generally work in these waters – even those seeking Sperm whales off the west coasts of South America worked not far from the land. Consequently the migration routes as mapped always appear to approach the coasts of the continents in their northbound course, and there is no information about what may be happening in the vast expanses of ocean far from the land.

Just as the whales of the southern hemisphere migrate yearly from the Antarctic towards the Equator, so those of the northern hemisphere migrate towards it from northern waters. It is obviously possible for whales to cross the Equator and join the population living in the other hemisphere, but marking returns have given no sign that this ever happens. Furthermore the probability that this might happen is reduced by the times of the migrations, for the whales go north or south from colder waters in the winter, so the migrations are out of phase by six months and a southern whale crossing the Equator during the southern winter would be unlikely to meet any whales belonging to the northern stocks, which at that time are at the northern end of their range.

Marking experiments have probably been most successful with the Humpback whale. They have shown not only the migration routes and times, but have also demonstrated that Humpbacks are divided into separate populations inhabiting different parts of the oceans, and that little mixing of the populations appears to occur. Observations on this species have also been comparatively easy because the animals come into inshore waters where they are more accessible to observation and exploitation; fisheries for breeding Humpbacks, for example, were carried on at Bahia de Todos os Santos on the tropical coast of Brazil, and on the west coasts of Australia.

Six distinct breeding grounds have been identified in warm waters:

119

on the Pacific and the Atlantic coasts of South America, on the Atlantic and Pacific coasts of Africa and on the east and west coasts of Australia. Sufficient marks have been recovered from Humpbacks to show that the populations of these breeding grounds are segregated, not only where they breed but also where they feed in the Antarctic and sub-Antarctic. Each feeding ground lies approximately south of the respective breeding grounds, and although each area has no definite limits the main concentrations of feeding Humpbacks are distinct. There is of course some scatter of individuals between the main concentrations, and probably a small amount of mixing between groups, so that whales born on one breeding ground sometimes find their way to the breeding ground of another group. Humpbacks, as we have already seen, sometimes at least feed while they are on migration, as in New Zealand where they feed on the abundant lobster krill. This is only to be expected, for it is not likely that any whale whether migrating or on its breeding grounds would refrain from having a meal at any time should it come across a supply of food.

In addition to the breeding grounds off the coasts of the continents there are also breeding grounds far out in the oceans though always near the shores of comparatively small islands. Thus many of the Humpbacks that pass on migration through New Zealand coastal waters go to breeding grounds in the neighbourhood of Tonga, Fiji and other Pacific islands [45]. Similarly in the Humpback populations of the northern hemisphere, which appear always to have been smaller than those of the southern, there is an annual migration from the feeding grounds in cold northern waters to the tropics and subtropics for breeding. The population of the western north Atlantic, which feeds in summer north of Newfoundland towards Greenland, goes south in winter to breeding grounds near the Bermudas and off the eastern Caribbean islands. The population of the eastern north Atlantic feeds off the coasts of Norway and beyond but moves south in winter to breeding grounds round the Cape Verde Islands and the adjacent coast of Africa. In the northern Pacific similar migrations take place between summer feeding grounds as far north as the Aleutian Islands and breeding grounds in the south China sea for the western population, and as far as the southern part of Lower California for the eastern group. Here again, although the breeding grounds are far apart there could be some exchange of individuals between groups on their northern feeding grounds.

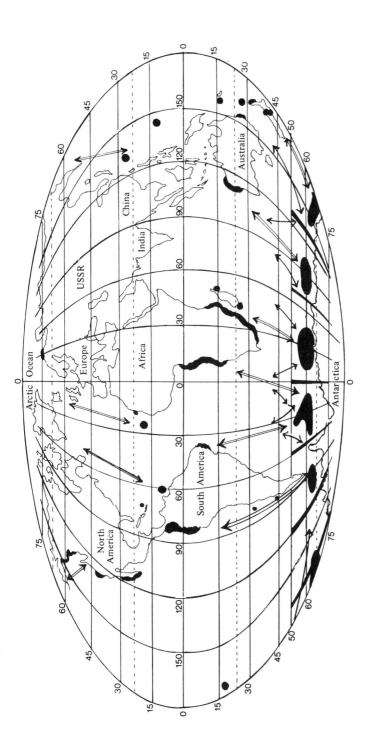

Figure 17. The migrations of the Humpback whale. The arrows show the approximate routes between the summer feeding grounds and the wintering grounds. The black areas show the positions of the grounds where Humpbacks were formerly hunted.

This picture of the migrations of the Humpback appears to be more clear-cut than it is in fact. It is in general correct without doubt, but the whales do not all migrate at the same time – the migrations are spread over a considerable period, and just as some individuals may stray away from their own group, so some may also not make a complete migration. This appears to apply more to the younger whales, which may linger in the warmer waters where they apparently find enough food to avoid the necessity of going to the Antarctic. We have seen that it is possible that whales may sometimes move across the equator from one hemisphere to the other, but there is another possibility for the mixing of northern and southern populations, as has been pointed out by Dr William Dawbin, the cetologist who has particularly studied the Humpbacks of the southern Pacific. Such a laggard in tropical waters could meet and breed with some of the northern whales on their southern migration; such occurrences would explain why the whales of the two hemispheres are the same species and have no structural or behavioural differences – they would share a common gene-pool. The same argument applies to the other baleen whales that have similar migrations between north and south.

The migrations of the Fin whale in the southern hemisphere are much less precisely known because, although there is a concentration of the animals on the southern feeding grounds, there is no similar concentration on coastal breeding grounds in warmer waters at the end of the northern migration. Even in the Antarctic the Fin whales are more widely dispersed, though naturally they were found in great numbers where the shoals of krill were most abundant, as off the coasts of South Georgia, in the neighbourhood of the South Shetlands, and in the Ross Sea, for example. The wide dispersal gives opportunity for greater lateral mixing between the populations that migrate northwards into the Atlantic and Pacific Oceans, but we do not know if this possibility is realized. There are no known concentrations of Fin whales on breeding grounds, which are probably widely spread over the open oceans far from land in temperate and subtropical latitudes. The scattered records of sightings of Fin whales from merchant ships show that during the southern winter the animals may be met with almost anywhere, from the coastal waters of Africa across the south Atlantic to those of South America, and similarly in the Pacific Ocean. The records from shore whaling stations indicate that Fin whales probably seldom go far across the tropic and do not habitually approach the Equator as closely as do the Humpbacks. The migrations are therefore much less channelled than those of the Humpback, and instead of a comparatively narrow stream of migrating whales there is a movement south or north over a wide front. Recoveries of marks from Fin whales in the Antarctic, however, do show that Fin whales tend to return to the same region of the Antarctic where they were originally marked [38, 167], although equally there is a definite tendency to wander to other parts with the passage of the years.

In the northern hemisphere there are similar seasonal migrations, but here again although there is a concentration of Fin whales on their northern feeding grounds, they are widely scattered on their southern breeding grounds. On their feeding grounds they tend to gather in those regions where there is an upwelling of water, as near the continental shelves, with a consequent rich production of potential food. There is, too, evidence that they find some food in the course of their migrations, and possibly even on their breeding grounds; but their main intake is when they are at their northern grounds, though it is only to be expected that they will feed whenever they may happen to find food elsewhere.

A recent line of research on the grouping of separate populations has been developed particularly by Japanese cetologists who by analysis of the blood serum of whales have shown that there are different identifiable blood groups among them [67]. In the north Pacific the Japanese have recognized by this means that there are different groups or sub-populations of Fin whales that appear to remain segregated. Extension of this research to the Antarctic has produced evidence of similar more or less closed populations.

The migrations of the Blue whale stocks of both hemispheres are similar to those of the Fin whale as far as is known, but it is probable that they feed even less when they are away from their cold water feeding grounds because they appear to be highly selective in their food and to rely almost exclusively for nourishment on the euphausiids found in colder waters. Unlike Fin whales, they do not take large quantities of fishes or other prey even when it is available, a matter for surprise considering that the Blue whale is the largest species, indeed it is the largest animal that has ever existed. One marking recovery from the southern population showed that the whale had moved to the opposite side of the Antarctic from where it was marked a few years previously [108], but a single record, though showing the possibility of population mixing, gives no indication about what habitually happens.

The Blue whale is now scarce, especially in the southern hemisphere, and is forbidden as quarry to the whaling industry, so further information on its migratory movements is not likely to be forthcoming in any quantity.

There is, however, a segregated population of Blue whales recognized by its smaller than usual average size and its restriction to feeding grounds in the sub-Antarctic. This is the so-called Pygmy Blue whale which frequents mainly the waters in the neighbourhood of Kergulen, or Desolation Island as it was formerly known, where it feeds mainly on a different species of euphausian, *E. vallentini*, from that forming the true Antarctic krill, *E. superba*. On its breeding migration it moves northwards into the Indian Ocean where like other stocks it disperses, though being a small distinct subpopulation it has a less wide distribution. Nevertheless it was once habitually taken late in the season in South Georgia waters, where the whalers knew it as the 'Mai-bjorn' Blue whale.

It is perhaps at first sight surprising that so little is known of the whereabouts of Fin and Blue whales when they leave their feeding grounds. On those grounds the concentration of the food into discrete areas leads to a similar concentration of the whales, but when the whales leave to travel north they disappear into the vast wastes of the oceans where they are dispersed beyond the possibility of observation. Even on their feeding grounds in the Antarctic in the days of their abundance they numbered no more than one whale to 20 or 30 square miles on average, so when widely scattered in warmer waters during the winter they just disappear into oceans without trace [107]. The understanding of the migrations of these species is further obscured by the fact that not all the whales of the stocks undertake the full migration every year; in particular the younger whales appear to remain north of the main feeding grounds where the concentration of krill is greatest. They are evidently able to find enough food for their needs in lower latitudes than those visited by the schools of the main part of the population.

The Sei whale, which feeds largely on planktonic crustaceans, goes as far as the ice edge in its feeding migrations in both hemispheres, but no segregation into separate populations has been found. In the North Atlantic, however, it is more commonly found in lower latitudes, and those years when it went to the far north in large numbers were unusual, and were known to whalers as 'Sei whale years'. Like the Blue and Fin whales it disperses widely in temperate and subtropical waters at the peak of the breeding migration, but it does not habitually penetrate into the tropics in any quantity.

On the other hand the entire population of Bryde's whale is confined to the warmer waters of all the oceans. Its feeding habits differ from those of the Sei whale, for it is not so extensive a feeder on planktonic Crustacea but eats fish and other comparatively large animals. Thus it appears to have no need to migrate to high latitudes on a feeding migration, and feeds and breeds in more or less a single area of distribution, though there is doubtless a considerable movement of individuals within it.

The Minke whale, though the smallest of the rorquals, penetrates as far north and south as the others, and goes deep into the pack ice. On its breeding migration it appears to be very widely distributed and less regular in the pattern of its journeys, for many remain in comparatively high latitudes during the winter when other species have mostly retreated to lower ones. The Norwegian cetologists have found that many Minke whales only a year or two old spend the winter in high latitudes as far north as the Barents Sea, and the Russians record the species in large concentrations of schools containing a thousand or more, in all parts of the Antarctic, though there is a tendency towards segregation into separate groups [8]. This much less clear-cut migratory behaviour when compared with that of the other rorquals may be correlated with the feeding habits of the species. It is much less dependent upon planktonic organisms and feeds extensively also upon quite large fishes such as cod and haddock, whereas the Fin whale when feeding on fish prefers smaller species such as herrings and capelin.

The migrations of the Gray whale, like those of the Humpback, are very much more stereotyped and consequently better known than those of the rorquals. The species is confined to the North Pacific where two populations both feed in the summer in high latitudes in the Sea of Okhotsk and the Bering Sea respectively. In the autumn the schools depart on their breeding migration towards the south, the western population going to the coastal waters of Korea, and the eastern one to those of California. There is no doubt about the whereabouts of the Gray whale in the breeding season, nor any question of its disappearing throughout the length and breadth of the oceans; its breeding grounds are close inshore where the animals can be watched with ease.

The movements of the eastern population are known best for the breeding grounds are the coastal waters and lagoons of Lower

California; the southern part of the migration route is so close inshore that the seasonal passage of the whale has become a tourist attraction on the western coast of the United States. These inshore habits and comparatively easy access to the animals when breeding led to such over-exploitation by commercial whalers during the nineteenth century that Gray whales were reduced to very low numbers, and the species was thought to be near extinction. The fishery came to an end for lack of quarry, but by the first quarter of the twentieth century numbers had become sufficient for it to be resumed for a few years, after which it declined again. Since then Gray whales have been given complete protection so that they are now once more plentiful. The Gray whale has shown a surprisingly quick restoration to reasonable abundance in the comparatively short period of complete protection, and may be an example of what can be expected if the populations of other over-fished species are left in peace to recover their numbers. The annual migration of the eastern population of the Gray whale covers a route 5,000 miles long, and the energy needed for travelling this great distance is taken in solely at its northern end, for the whales do not feed at all on their breeding grounds nor on the southern parts of their journeys.

The Right whales are the least migratory of the mysticetes, but they have been little studied in modern times because they have had no commercial importance since the great reduction in their numbers brought about by over-fishing in the nineteenth century. The Greenland whale, circumpolar in its distribution, and confined to the northern hemisphere, is almost sedentary, for it remains in the neighbourhood of the northern ice throughout the year, and its southwards movement in winter is determined by the position of the ice edge. The Black Right whales of both hemispheres are inhabitants of the temperate and sub-boreal seas, and although there is some seasonal movement north and south they do not undertake lengthy and distinct feeding and breeding migrations like those of the rorquals. The females, however, appear to have favoured coastal waters and the shelter of bays for the birth of the young. Again, there is no information on whether they feed during the breeding season, though there appears to be no reason for refraining from food whenever it is available. The species has not increased so quickly as has the Gray whale on the ceasing of commercial exploitation, though it too may be expected to do so in the course of time.

Among the odontocetes, the Sperm whale is the only species that has been subjected to a worldwide fishery of long standing, and is consequently the species about whose movements we know most. The Sperm whale is essentially a warm and temperate water species and its food, unlike that of the rorquals, is not concentrated in any particular part of the world by the high seasonal production of the polar waters. There is consequently no cause for it to make regular long migrations between feeding and breeding grounds. It is cosmopolitan in its distribution, and can be found at all times of the year in the tropics of the Atlantic, Indian and Pacific Oceans. Nevertheless there is some indication that there is a drift towards the south or north during the summers of each hemisphere. This was shown in the distribution maps prepared by the American cetologist Townsend, who plotted the positions of capture of great numbers of Sperm whales by examining the log-books of the old sailing whalers [202].

In addition quite large numbers, almost entirely of bull Sperm whales, find their way into high latitudes of the Arctic and Antarctic in the summer months of the respective hemispheres. These animals are not, of course, feeding on the abundant plankton of the regions, but they may well be there indirectly as a result of its abundance, for they feed upon the fishes and squids that prey upon it. The cows are not found in any numbers in high latitudes, so that the calves are born in temperate and tropical seas. Sperm whales often occur in schools and thus show a tendency towards segregation into separate groups, but though this does no doubt point towards there being local populations there is no sign that they differ even subspecifically in the different oceans or parts of them. The bulls at least are sometimes great wanderers and there is no reason why bulls visiting the Antarctic from one of the great oceans should not transfer to one of the others when they turn northwards as summer ends.

Similar considerations apply to many of the other odontocetes. They are all predators on squids and fishes, and consequently can only be indirectly affected by the distribution of planktonic animals. Furthermore, owing to the lack of any great commercial value of most species, they are not captured in numbers large enough to give any precise information about their movements. The exceptions are various local fisheries for small odontocetes, of which the most important from this point of view are those of Japan, where porpoise and dolphin meat is highly esteemed and where moreover there is an energetic and well-staffed Whales Research Institute.

Some of the species are cosmopolitan in their distribution, but if they are not particularly abundant anywhere the opportunities for observing them are rare. The identification of such species usually depends upon the finding of animals accidentally stranded; when one considers the vast extent of the world's coasts where it is unlikely that they will be found by anyone, let alone by a cetologist competent to identify and record them, the dearth of information is not surprising. A good example is provided by Fraser's dolphin, *Lagenodelphis hosei*, which was known from a single skeleton collected in Borneo in 1895 but not recognized as a distinct species until 1956, yet since 1970 has been found in the Philippines, Taiwan, Japan, and other parts of the Pacific, eastern Australia and South Africa.

From direct observation, from the records of accidental strandings, and from records of the catches at local fisheries for dolphins and porpoises, it is evident that many species of the smaller odontocetes make seasonal movements, which are not, however, as extensive or so regular as those of the mysticetes. Such movements depend upon changing oceanographical conditions, and are often correlated with the temperature of the water, with the variations in the position of upwellings near continental shelves, and with the consequent abundance or dearth of prey animals. Some of the fisheries for odontocetes are seasonal and hence show that the animals are captured in the course of more or less regular seasonal movements.

One such was the fishery for the Common porpoise in the Little Belt between Funen and Jutland in Denmark. The ice forming in the northern part of the Baltic drove the porpoises out into the North Sea, and those that took the route through the Little Belt were netted in November and December. This annual migration came to an end in the 1960s owing to the drastic decline in the number of Common porpoises in the Baltic, probably caused by the pollution of that sea, though whether acting directly on the porpoises or indirectly through their food is not known. A similar fall in the number of Common porpoises has been noticed in the North Sea, so its cause is not confined to the Baltic.

Some of the seasonal movements appear to be onshore and offshore rather than lengthy north to south migrations; those of the Common porpoise generally follow this pattern, and probably correspond with similar movements of fish and particularly squid. Among the ziphiids the Bottle-nosed whale, *Hyperoodon ampullatus*, in the north Atlantic shows a more well defined migration than many other odontocetes, for it travels north as far as the edge of the ice in summer and south as far as the Cape Verde Islands in the winter. There is, however, evidence that there may be both migratory and non-migratory groups, for in some places they are present all the year round. The White whale, *Delphinapterus leucas*, too, makes regular seasonal movements, though of lesser extent. In the summer large schools of this species come into shallow water and enter the estuaries of rivers on the Atlantic coasts of North America as far south as the St Lawrence where, however, the local population appears to be a separate one from that occurring farther north. This inshore migration is a breeding migration – the young are born in shallow water where the schools are concentrated into large aggregations which make an impressive sight when seen from the air.

The Pilot whales of both species of the genus *Globicephala* undertake considerable but irregular seasonal movements following shoals of fish and squid. Here again the movements though irregular do show some seasonal correlation, for the ancient fisheries in which schools of the whales are driven ashore and captured by traditional methods take place mostly in the summer months in Faroe, as they did formerly in Orkney. Similar fisheries for this species in Newfoundland show a corresponding irregularly seasonal incidence. The Bottle-nosed dolphin, *Tursiops truncatus*, the favourite of the marine aquaria, has long been known to make coastal movements both alongshore and on- and off-shore along the Atlantic coast of America, where there was for long a commercial fishery at Cape Hatteras, but the details of the movements and the factors affecting them are complex and are imperfectly understood.

An extensive fishery that now takes large numbers of porpoises and dolphins has been developed in the tropical Atlantic and subtropical regions of both the Atlantic and Pacific Oceans. This is the purse-seine fishery for tuna fish, which occurs in immense shoals with which associate large schools of dolphins for reasons which are obscure because they are not mainly in pursuit of the same prey. The fisheries are carried on in oceanic waters up to several hundred miles from shore by ships of several nationalities, some from a great distance. The main quarry is the tuna, but great numbers of small odontocetes of several species are taken incidentally. In the Pacific as many as 350,000 dolphins have been killed annually and their bodies thrown away without being put to any useful purpose. Methods are now being explored for finding a way to make commercial use of this great waste of potentially valuable material, or, alternatively, to avoid killing the dolphins. In the Atlantic the ships operating from west African ports also catch great numbers of dolphins in the tuna nets, but here they are not wasted for they are esteemed as food in those regions and consequently are properly utilized. Several species are caught in the tuna fisheries; in the Pacific one of the most commonly taken is the Spinner dolphin, *Stenella longirostris*, which makes such a spectacular display when large shoals breach from the water with the animals spinning on their long axes before falling back into the sea. These great catches of dolphins do not appear to have thrown much light on any possible migrations of the species concerned; tuna aggregate not only round schools of dolphins but also round objects such as pieces of driftwood, a habit also exploited by the fishery.

In a more restricted area of the oceans the seas around the shores of the British Isles have yielded some information about the seasonal movements of odontocetes. In 1913 the Director of the British Museum (Natural History), the late Sir Sidney Harmer, came to an arrangement with the Coast Guard and the Receiver of Wreck, that all cetaceans found stranded on the British coasts should be reported by telegram to the Museum, and that the animals should be reserved for the Museum's use if needed. This arrangement has now continued for over sixty years, and although the reports for any one year may seem sporadic and scattered, the accumulated information now makes it possible to recognize seasonal movements in many species of cetaceans that regularly inhabit the British seas. Agreement with the authorities, which now includes the Customs Service, was reached the more easily because for hundreds of years stranded cetaceans – and captured sturgeons – were in England designated 'royal fish' and were the property of the Crown. The legislation remained in being although it had fallen into disuse in modern times; when in recent years this and other obsolete laws were being 'tidied up', certain British cetologists were able to persuade the legislature to retain the law relating to royal fish, so that the reporting scheme could continue, much to the advantage of cetology.

Many of the species found are rare and consequently the records are scanty for them, though they give valuable information about their general distribution; some are evidently near the limits of their distribution in British waters. The more common species, on the other hand, give so many records that taken together they do indicate the pattern of their usual movements. For the Common porpoise, *Phocoena phocoena*, an analysis of the occurrences shows that strandings are most frequent during the period July to October, and that there is another spate of strandings in January to March in the North Sea. It is known from the records of a Danish seasonal fishery that there is an annual migration of the species out of the Baltic into the North Sea between November and February, and this is probably correlated with the subsequent occurrence of porpoises on the east coasts of England during the summer. Porpoises evidently enter the northern part of the North Sea and work southward to the English Channel. It is improbable, however, that all the porpoises in the British seas, where they are the commonest cetacean, migrate from the Baltic. The species is common in the waters of both sides of the Atlantic, and an annual migration brings it in large numbers to British waters during the summer. It is not clear, however, where the majority come from nor where they go, apart from the comparatively small population that spends the summer in the Baltic.

The Common dolphin, *Delphinus delphis*, inhabits the warmer parts of the Atlantic and the Mediterranean, and comes as far north as the British Isles. It is particularly common in the English Channel, the southern part of the Irish Sea, and off the Irish coasts. It is less frequent in Scotch waters and is usually rare in the North Sea. In 1933 and again in 1937, as we have seen, an exceptional influx of Atlantic water came into the North Sea round the north of Scotland, bringing with it large numbers of animals not usually found there, including great quantities of the squid *Todarodes sagittatus*. Large shoals of the Common dolphin came with this influx in pursuit of the squid, and those stranded on the east coasts of Scotland and England showed that they had been feeding on the squid by the presence of numerous marks made by the suckers of the squids on their jaws.

The White-sided dolphin, *Lagenorhynchus acutus*, is numerous in northern Scotch waters, where it often occurs in extremely large schools numbering many hundreds, perhaps thousands, but as its main area of distribution is to the north, the stranding records have not indicated any migratory movements along the British coasts. The White-beaked dolphin, *Lagenorhynchus albirostris*, on the other hand, is less gregarious and is generally seen in small schools numbering twenty to thirty. It is abundant in the North Sea where it migrates northwards during the summer and returns southwards in the autumn and winter. It is also seen in lesser numbers on the west coasts of England, Scotland and Ireland, but records are not sufficient to show any migratory movements; it is rarely seen in the English Channel between the North Sea and the west coasts.

Although the Bottle-nosed dolphin, *Tursiops truncatus*, has been seen in British seas at all times of the year, there is a great increase in its numbers during the summer. An annual migration approaches from the south-west in early summer, and the dolphins spread up the English Channel, the Bristol Channel and St George's Channel, and up the west coast of Ireland. Some of them pass through the Straits of Dover but they do not go far into the North Sea, as few get beyond the coasts of Suffolk and Holland; on the west it is rare off the Atlantic coast of Scotland. The animals first appear in some numbers in British waters in May, and reach their greatest abundance in August, after which they become scarcer. The autumn migration away from the coasts is believed to be in the direction opposite to that of the early summer; but the destination of this south-easterly journey is not known, possibly it is away from European coastal waters into the more open Atlantic.

The information about the movements of other odontocetes is less complete, but the occurrences of some of the less common species, though scanty, throw some light on their seasonal travels. For example Risso's dolphin, *Grampus griseus*, a species not often seen in British waters, has usually been found on the south-west coast of England and to the south and west of Ireland. Its main area of distribution evidently lies to the south-west, as far as those that come into British waters are concerned. It sometimes strays as far north as Scotland and very rarely into the North Sea, but although odd specimens have turned up in most months of the year, it is evident that the annual approach to inshore waters takes place in the late summer. It is of interest, too, to notice that the Sperm whale, which is generally an uncommon straggler to British coasts, turned up on both sides of the North Sea in 1937, the year when the unusual influx of large squid attracted great numbers of Common dolphin. It is probable that the Sperm whales were also pursuing and feeding upon these squid. The records of stranded whales on British coasts also contain a number of instances of the mass stranding of schools of several species of odontocetes, a subject discussed further in Chapter 7.

Most of the mysticetes are merely strays into British coastal waters, with the exception of the Minke whale, *Balaenoptera acutorostrata*, which has frequently been stranded. The majority of this species has been found in August, though odd examples have been found in every month, and a lesser concentration has been recorded in the early summer. It appears that there is a northward migration towards the Norwegian coast in spring and a return southwards in autumn. The Minke whale rarely goes up the English Channel, or farther south than the Wash in the North Sea. Some of the animals must therefore return northwards after their visit to the North Sea, and others evidently join any southward stream west of the British Isles, if indeed there is one, for we have already seen that Minke whales have less well marked migrations than the other rorquals, and often winter in quite high northern latitudes.

The other migratory rorquals, the Humpback, and the Sperm whale, regularly pass the British Isles on their annual migrations, but at a distance. In the early decades of the present century several shore whaling stations worked in British waters on the west coast of Ireland, the Scotch Hebrides and Shetland. They caught their whales about a hundred miles offshore in the neighbourhood of the edge of the continental shelf at the hundred fathom line, which forms the boundary of the warm ocean current. The seasonal peaks in whaling activity at these stations show that the whales were taken in the course of their southern and northern migrations, and supply only one detail of the broad picture of them that has been drawn from the whaling returns from much wider areas of the globe.

The results of the British Museum's stranded whale reporting scheme well demonstrate how the patient accumulation of information over a large number of years gradually builds up to give a reliable account of the movements of Cetacea. They have also provided the systematists with much valuable material that elucidates the characters, anatomy and other points in the biology of those species that are rarely seen, and are otherwise inaccessible because they have no commercial value and consequently are not deliberately captured by man.

We have seen that the whales that undertake long and regular migrations are mostly the mysticetes, for the movements of the odontocetes appear to be less well defined because their food is more widely spread and not concentrated into specific regions. Whalers have for long known that the whales that first appear every season on the feeding grounds, especially in the Antarctic, are lean, whereas after a stay of several months during which there is abundant feeding, they are fat. The comparative oil yields from whales at different times in the whaling season are a simple demonstration of this fact.

133

Some species, such as the Blue and Gray whales, seem to concentrate the greater part of their feeding into the time spent at the polar ends of their migrations among the rich plankton of the colder seas. They feed very little at other times, though Gray whales have been recorded as feeding on fishes during their southern migration, and some Blue whales were said by Scammon [175] to feed on sardines and shrimps off the coast of California, and others are thought to feed in the seas south and east of Madagascar. On the whole, however, these species feed mainly if not entirely on the 'feeding grounds', and fast for the greater part of the time they are away from them. The question then arises, how do whales carry sufficient fuel to produce the energy needed for taking their long journeys?

In the first place their rate of metabolism seems to be much slower than that of such frisky animals as the smaller odontocetes; their rate of respiration is slower, showing that the burning of fuel is less rapid. The Japanese cetologist Kawamura has made some interesting calculations based on an examination of the blubber of whales in the Antarctic [98]. He concluded that the whales use up to about 2 centimetres of the thickness of their blubber in a five-month migration from south to north and back; in a 60 foot Fin whale the oil content of this blubber thickness is about 1.8 cubic metres, yielding 9.34 to 9.45 kilocalories per gram. Thus a whale of this length would have about 0.36 cubic metres of available oil for use per month in locomotion, which corresponds to over 15 million kilocalories for a there-and-back migration lasting five months. If the whale swims at 10 knots, and the flow of water over its body is laminar, this would provide energy enough for covering about 20,000 kilometres.

It is, however, unlikely that whales follow a straight course on their migrations; they probably alter course, interrupt their migration in an irregular manner, even stopping to 'play around'. Consequently the total distance travelled in the course of a five-month migration could be 28,000 to 30,000 kilometres, so that the whales would need to eat some food on the way, as indeed we know they do. Kawamura also suggests that whales do not perhaps travel so far as has been thought, but that they breed in waters of higher latitude than the temperate and tropical seas that are generally accepted as the breeding grounds. Against this suggestion the known sightings of whales in seas of lower latitude must be set. As Kawamura admits, we really do not know where the great whales go in the winter season, and we have little information about the daily lives of whales, or how much time they spend 'playing around' instead of swimming on their course without interruption. We do know that even on their feeding grounds whales often move about for considerable distances, up to 50 miles or more, from day to day, and thus there is no reason to suppose that they follow a very definite route when they migrate.

There is for most species not any clearly defined 'corridor' along which the animals move, such as is seen in many kinds of migrating birds. The whole migration of whales is a much more diffuse movement of generally scattered individuals than the dense flocks of some migratory birds or shoals of fish. Nevertheless whales do have the possibility of making a round trip corresponding to 120 degrees or more of latitude, and consequently the possibility of crossing the Equator, so there could be occasional transfers of individuals from the southern to the northern populations.

Kawamura's suggestion that the conception of a migration to warm temperate and tropical seas for breeding is not proven appears to be itself refuted through the records of sightings of whales in these regions and by the known fact that most, perhaps all, species do feed to some extent in these areas and during the course of their migrations. It would indeed be surprising if whales did not feed whenever they have the opportunity during their wanderings, apart from the annual feast on the definite feeding grounds in the colder seas. We may therefore see the migratory patterns of the mysticetes as a general and diffuse movement from north to south and vice versa, during which the animals are not necessarily forced to fast, but may delay their journey to feed wherever they may find anything palatable to eat, just as any animal would when it is hungry.

Although something has been discovered about the migrations of the cetaceans, nothing is known about how they, or indeed any other migratory animals, carry them out. The matter is particularly puzzling in considering cetaceans, and oceanic birds, which migrate over immense distances of featureless ocean with no landmarks for guidance. It is true that some species, especially the Gray whale and the Humpback, follow the coastlines in part at least of their annual journeys, but they do not remain entirely in coastal waters. The Humpback in its migrations from the Antarctic, for example, has to cross the width of the southern ocean before it makes contact with the shores of the continents or of New Zealand. Similarly the Gray whale, at the northern end of its migratory path in the east Pacific, leaves the neighbourhood of the American coast and makes for the Aleutian chain of islands and the Bering Sea. Whales do sometimes poke their heads out of the water so that they may see things above the surface, so it is possible that these species do see the coastline, and perhaps they recognize landmarks in that way.

It is also possible that the odontocetes may become aware of neighbouring shores, and of shallowing water near them, by echolocation, but here again awareness of the proximity of land cannot imply any recognition of landmarks, not can it aid in setting a course when traversing the open ocean. Echolocation cannot be suggested as an aid to navigation for the mysticetes. Direct sound signals might be a possible method of following those that have gone ahead, because the sounds emitted by cetaceans travel to great distances under water – but what of those at the front who have none before them to give signals? Furthermore such a system of following could not apply to most of the mysticetes, which do not follow narrow migration routes but disperse over great areas.

It has been suggested that migratory animals possess a way of navigation that depends, like the navigation of ships by man, on a knowledge of position defined by latitude and longitude. For this it is assumed that the animals find their approximate latitude by observing the meridian altitude of the sun, and find their longitude by combining this observation with information from an 'internal clock', perhaps some phenomenon connected with the 'circadial' or daily metabolic rhythm. The advocates of this theory overlook the fact that to a castaway in a lifeboat in the open ocean no quantity of sextants or of chronometers are of the slightest use unless he has a chart – supposing he can manage without the nautical almanac. Furthermore anyone in such a predicament must know not only where he is, but whither he wants to go. It is inconceivable that cetaceans, or any other animals, have even the crudest sort of mental chart of any part of the oceans, and it is equally improbable that they can have any knowledge of where they are going or why.

Migration is part of the built-in behaviour of the animals and is in some way incorporated in the genetic code, just as are the behaviour patterns that give birds characteristic songs or make them build nests of characteristic shape, size and particular materials. But even an innate behaviour pattern needs some stimulus, internal or external, to start its expression; there may be internal stimuli such as the production of sex hormones, or the degree of fatness after a season of feeding or fasting. There is, however, an external stimulus that could well be the trigger – the increase or decrease in the length of the day.

How behaviour patterns can be transmitted in the genetic code which consists of sequences of bases in the molecule of deoxyribosenucleic acid, has not yet been explained. It is, however, possible that a code which specifies for example 'build a Blue whale' makes a creature that not only looks like a Blue whale but automatically behaves like one. The innate behaviour patterns of animals are as much part of their specific characters as their anatomical structure.

Chapter 7

Communication and echolocation

It is strange that until the end of the first half of the present century zoologists held that whales had a poor sense of hearing and that the different parts of the auditory apparatus did not function in transmitting sound to the inner ear. Some held that cetaceans were entirely deaf to waterborne sounds, others that such sounds were perceived, not through the mediation of the outer and middle ear structures but by direct conduction through the bones of the skull to the cochlea.

These misconceptions are the more surprising because the anatomy of the hearing organs in cetaceans had been described in detail by many cetologists. At the end of the eighteenth century John Hunter had made researches on the subject, and in 1868 Richard Owen published a precise account of the anatomy of the auditory tract in Cetacea and noted that the auditory nerve is large. He compared the closed external meatus with that of man if filled with water, noting that a drop of water falling on that contained in a water-filled human meatus produces 'a sensible impression of sound'. He concluded that 'the Cetacea, with their meatus and ear-drum in a like condition, would thus be affected by any sonorous vibrations that might be propagated to the tympanic cavity' – that water-borne sounds would be transmitted to the cochlea as in other mammals and would consequently be heard perfectly [143].

Had those who persisted in upholding their erroneous theories troubled to think of the behaviour of living cetaceans they might have more readily accepted that animals provided with elaborate organs of hearing might be expected to use them. Since men first took to whaling they have known that cetaceans have an acute sense of hearing, and that nothing is more likely to frighten away a whale that they try to approach than a sudden sound inadvertently made in the whale boat. In modern whaling with powered catcher ships great care was taken to prevent any noise being made that might spoil the chase – a bucket accidentally dropped on deck as the approach became close could scare the whale and ruin the chances of a shot. In the large catchers now used, however, the technique is reversed; supersonic sounds are deliberately broadcast underwater with an 'ultrasound gun' to keep the quarry on the run while followed by the high-powered catcher until exhaustion brings it to the surface and makes it an easy prey, or to drive it within range of another catcher.

Another possible reason why so many persisted in their perverse belief in the face of evidence to the contrary is that whales do not have a voice similar to that of other mammals. They do not emit sounds by expelling a stream of vibrating air from the mouth, an impossibility in a submerged cetacean, and they lack any vocal cords in the larynx, as noted long ago by Owen and others. That does not mean, however, that they cannot produce any sound, far from it; those who believed them to be dumb forgot that the wind side of the orchestra produces plenty of sound from instruments other than those of the reed family such as the clarinet and oboe – the cetacean larynx is extremely complicated and quite able to produce sounds. Furthermore had they considered the living animal they might have discovered that whalers have long been familiar with the sounds emitted by at least some Cetacea. The Arctic whalers of the nineteenth century, who travelled silently under sail and used the auxiliary power of noisy steam engines only when necessary, knew the liquid trill of White whales so well that they nicknamed the animals 'sea canaries'.

It was not until 1960 that the elaborate nonsense proclaimed by those who allowed their preconceived notions to blind them to the evidence given by their own eyes and ears was finally shattered by the work of two British cetologists, F. C. Fraser and P. E. Purves, who, not content with their own most painstaking and careful anatomical investigations, went to the trouble of devising some elaborate experiments to find out how the structures actually work [66]. Their results swept away the mass of misunderstanding and gave a clear scientific account of the whole subject of hearing in the cetaceans – one cannot help being reminded of the story of the unprejudiced child and the emperor's new clothes.

Their researches began with an investigation of the elaborate system of air sinuses, or dilations of the respiratory passages, characteristic of cetaceans, that had long been known to, and had as long puzzled, zoologists from the time of John Hunter. Such diverticula of the breath passages are present in many mammals, but in the Cetacea they are particularly complicated and difficult to study, for they are filled with foam, and so inextricably surrounded by and embedded in masses of blood vessels and fatty tissue that their definition by straight dissection is practically impossible. The only way by which their structure can be revealed is by injecting the sinuses, the arteries and the plexuses of veins with different coloured substances that solidify after filling them, so that the arrangement can be seen after dissection with the knife or with corrosive fluids.

The air sinuses commonly found in the bones of the face in most mammals, such as those in our frontal bone above the eyes and those in our cheekbones, are not present in cetaceans. These facial sinuses of the bones are connected with the nasal passages, but the sinuses of the cetaceans are connected with, or extensions of, the cavity of the middle ear which lies inside the ear drum and contains the chain of small bones that carry vibrations from the drum to the cochlea in the inner ear. Cetaceans do have sinuses connected with the nasal passages, but they are not contained within the bones of the facial part of the skull; they form a complex series of air cavities extending through the soft tissues outside the skull below the blow-hole.

The middle ear cavity is connected to the pharynx by the Eustachian tube, which serves in land mammals to equalize the pressure on each side of the drum and thus prevents its being strained or burst by variations of air pressure. In cetaceans the cavity of the middle ear is in effect ballooned out so that extensions of it have invaded the tissues and bones of the basal part of the skull, and insinuated branches between, and formed envelopes round, various organs, making a system of interconnecting air spaces of great complexity, which differ in their details from species to species yet form a series that is suggestive of the way in which the sinuses have evolved.

Five main divisions of the sinuses are recognized: an anterior sinus running forwards in the rostrum beneath the upper jaw, a pterygoid sinus forming a large cavity from which the anterior sinus originates; a peribullary sinus, also communicating with the anterior sinus, containing the ear bones and partly lining the tympanic bulla commonly called the 'ear bone'; a posterior sinus connected with the tympanic cavity and deeply embedded in fat; and a medial sinus near the jaw joint. All the sinuses are divided up into complicated air pockets by internal strands of tissue, and surrounded for the greater part by large and complex networks of arteries and veins. In the rorquals and the Humpback the cavity of the 'glove finger' extending into the external ear passage from the tympanic cavity is similarly an extension of the lining of the middle ear cavity.

The great complexity of the variations so carefully elucidated by Fraser and Purves is beyond the scope of this account, but some of the unusual points about them must be mentioned. The pterygoid sinus gives off diverticula which surround the optic nerve; the Eustachian tube, which runs in a winding course from the middle ear to the nasal cavity, is closely applied to part of it. Its most outstanding feature, however, is its relationship to the pterygoid and neighbouring bones. The Cetacea provide a progressive series from the Right whales to the Common dolphin, showing an increasing invasion and erosion of the pterygoid bone by the sinus. In the simplest example the sinus bulges into a hollow in the pterygoid, and as the series goes on the invasion gets greater so that in the rorquals the pterygoid is effectively split into two laminae separated by the sinus. Farther on in the series the sinus extends and erodes the inner layer of the pterygoid until it completely disappears, leaving the sinus in contact with the alisphenoid bone above and the remains of the lower layer of the pterygoid below. A series of developments in a lateral direction take place at the same time, giving an increasing complexity of diverticula among the bones of the skull and the surrounding tissues. As all these complications take place in three dimensions it is difficult to understand their arrangement and interrelations. On the other hand, the presence and invasive nature of the sinuses go far to explain the peculiar features of the bones of the cetacean skull, and it is fair to say that without a knowledge of them the cetologist is at a loss in understanding the morphology and architecture of the bones.

The question immediately arises, what use do the sinuses serve in the physiology of cetaceans, and how do they come to modify the apparently rigid structure of the bones? As all the essential mechanisms for mammalian hearing are present in cetaceans – meatus, drum, ossicles, and inner ear, though modified from the usual shapes, it is necessary to maintain an air space in the middle ear; the sinuses and their contents serve to regulate the pressure in the middle ear when the animals are under great pressure during diving. In land mammals the equalization of pressure on each side of the drum is automatic through the passage of air through the Eustachian tube under atmospheric pressure, but for a

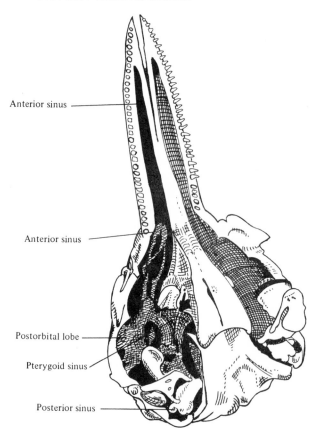

Anterior sinus

Anterior sinus

Postorbital lobe

Pterygoid sinus

Posterior sinus

Figure 18. The under side of the skull of a Common dolphin showing the positions of the sinuses, which have been injected with latex. (Modified and simplified from Fraser and Purves.)

cetacean at depth the pressure is enormously greater, and the pterygoid air sinuses form a reservoir for regulating it.

As already mentioned, the sinuses are filled with foam consisting of an oil–mucus emulsion, so the compressibility of the sinuses depends upon the compressibility of the foam. Fraser and Purves therefore carried out experiments with different sorts of foam to find the degree of persistence of the gas in the foam under great pressure, the persistence being necessary to maintain acoustic isolation under such conditions. They prepared foams with albuminous and with gelatinous bases and observed the behaviour of the gas bubbles as the pressure was increased, using a microscope and a transparent pressure cell. As the pressure was increased the volume of the foam decreased and the foam became a mass of minute bubbles, separated by distances about equal to their diameter, dispersed in the liquid medium. Both kinds of foam behaved similarly and were stable at pressures up to 100 atmospheres, equivalent to a depth of over 3,000 feet. When the pressure was released the foam structure reappeared at atmospheric pressure. On the other hand a foam made from domestic detergent collapsed under pressure and showed no minute bubbles. These experiments show that gas bubbles would persist in the sinuses even at the greatest depths to which cetaceans normally go, so that a sound-reflecting system surrounds the essential parts of the ear.

The extensive air sinuses are bound to decrease in volume as pressure increases, and here the function of the plexuses of blood vessels comes into play. When the volume of the sinuses decreases the plexuses become correspondingly distended with blood and act as space fillers so that there is no disturbance or breakdown of the surrounding structures. These arrangements ensure that there is always an air space in the cavity of the middle ear whatever the pressure, and the tympanic bulla is saved from fracture by the engorgement of a body of cavernous tissue with blood from the internal carotid artery, thus filling the tympanic cavity enough to preserve the bulla from damage.

The foam in the sinuses is produced by the secretions from the network of crypts and mucus ducts that line them, together with the air in them. The tympanic cavity, however, lacks such secretory glands, so that although the sinuses are filled with foam, the tympanic bulla remains filled with gas, which at extreme pressure would surround only the chain of auditory ossicles.

It is of interest to note that, although the attenuation of bone in the skull impinged upon by the sinuses is genetically determined, the parts concerned are exactly those that would be eroded when subjected to pressure. Bone, although it is a rigid part of the body, is more quickly modified than the soft parts when subjected to stresses, and in particular it is resorbed in areas subjected to unusual pressure, just as the comparatively soft tyre of a motor car wears the hard surface of the road more than it wears away its own tread. On the other hand, the gas-filled bulla, in spite of the space-filling cavernous tissue, is so massive and thick that it can sustain some amount of pressure difference while the internal adjustment of pressure is made.

143

The demonstration of the structure and mode of action of this elaborate system that ensures the acoustic insulation of the middle ear, so that vibrations are transferred to the inner ear even at the greatest pressures met with in diving, and protects the whole auditory apparatus from pressure damage, proves that cetaceans, far from being defective in hearing, possess the means for perfect hearing under conditions to which no other mammals are ever subjected.

The foam-filled sinuses, however, probably have an additional though perhaps subsidiary function. Approximately four-fifths of the air in the sinuses consists of nitrogen, and we have already seen that the dissolving of nitrogen in the blood can be a danger to animals returning to atmospheric from greater pressures. Will not, therefore, the nitrogen in the sinuses dissolve in the blood during a dive and be released from solution as dangerous bubbles on returning to the surface? Fraser and Purves point out that the gas in the sinuses is not free but is contained in small bubbles in an oil–mucus emulsion. The solubility of nitrogen is very much greater in oil than in a watery medium such as blood, and consequently the nitrogen is absorbed into the oil in preference to dissolving in the blood. This process is enhanced by the very large total surface of oil presented to the gas by the combined areas of the surfaces of innumerable minute bubbles – Fraser and Purves mention that 100 millilitres of oil – less than $\frac{1}{8}$ pint – broken down into particles one-thousandth of a millimetre in diameter would have a total surface area of 1,200 square metres – larger than a tennis court. When pressure decreases nitrogen is released into the foam to increase its volume, but the rate of release is slower than the decompression rate if a whale is surfacing rapidly. They conclude that there would be an accumulation of nitrogen in the foam and that 'there is evidence that this nitrogen foam is blown out at expiration', as we have seen in Chapter 5. A word of caution is, however, needed; it has been suggested that the foam is a post-mortem artifact, and not present during life.

Coming now to the mechanism for hearing that is so elaborately protected by the sinus system we find that all the usual mammalian components are present, but that they all differ in detail in correlation with the reception of underwater sounds. In land mammals the external ear is generally a funnel-shaped pinna, often of comparatively large size, that directs sound waves into the meatus or passage leading into the ear drum. The cavity of the middle ear lies on the other side of the drum and contains three small bones linked together, the first attached to the drum and the third to the oval window of the inner ear, which consists of the coiled tubular cochlea in which lie the nerve endings stimulated by the pressure waves of sound. The bones form an impedance matching transformer with automatic volume control, and are so arranged that though the degree of movement transmitted to the inner ear is less than that at the drum, the pressure is increased. Thus in land mammals the hearing apparatus is such that it is an amplifier, and is highly sensitive to faint sounds. In water, on the other hand, the pressure amplitude of sound waves is sixty times greater than that of sounds of the same intensity and frequency in air; cetaceans therefore live in a world full of noise much louder than that of land mammals, so their hearing apparatus, far from amplifying sounds received, serves to attenuate them. Yet it is essential that cetaceans should be able to hear the sounds relevant to their lives at a threshold level similar to that enjoyed by land mammals – the whole of the cetacean auditory tract ensures exactly this, at the same time protecting it from damage through excessive noise.

As we have seen, the outer opening of the ear is minute, and even in the largest whales is no more than a centimetre in greatest diameter. There is no external ear pinna or conch, but it is interesting to note that for part of its course the meatus is wrapped round with a cartilaginous body that is in effect the ear pinna withdrawn below the surface – there are even some rudimentary muscles that represent those that move the pinna in land mammals. In the odontocetes the meatus, though minute in diameter, runs from the surface to the drum, though in some specimens of the Sperm whale it has been found to consist of discontinuous sections of tube. In the mysticetes, on the other hand, the meatus is closed for a distance of several inches, and then opens out into the part formed by the ear plug which is attached at its base to the glove finger described in Chapter 4. The ear plug is a good transmitter of sound vibrations and forms an essential part of the path taken in conducting vibrations inwards. The shape and arrangement of the middle ear bones is, however, such that the displacement amplitude of the innermost ear bone by a waterborne sound is similar to that of the same sound if airborne (figs 12, 13).

The experiments made by Fraser and Purves show that the sound attenuation of the tissues and structures surrounding the hearing apparatus is much greater than that of the meatus, and consequently cetaceans are not deprived of the advantage of binaural hearing and its

145

use for the perception of the direction from which sounds emanate. Land mammals are able to perceive the direction from which sounds come either by comparison of the intensity of sound received by the two ears or by moving the head so that the intensity is the same in each ear. In cetaceans, too, the meatus, though comparatively much less in diameter in order to avoid reception of excessive stimulation, is the most favourable path for the transmission of sound. Consequently it provides a pathway on each side of the head so that perception of direction is attained exactly as it is in land mammals. The apparent rudimentary or even degenerate condition of the auditory apparatus is entirely illusory and, far from being of little use, is a highly efficient mechanism that copes with the particular difficulties met with by animals living in an acoustic environment unlike that of animals surrounded by air. It gives acute hearing in spite of the high level of extraneous noise, and at the same time protects the delicate mechanism of the middle and inner ear from the effects of pressures unknown to land mammals.

There are further anatomical and functional details, too complex for discussion here, that show the correctness of the conclusions reached by Fraser and Purves, such as the rotational movement of the malleus being the only one that can be transmitted to the incus, thus preventing resonant vibrations from the tympanic bulla or other bones of the skull from being transmitted to the cochlea.

These admirable researches give us a very exact understanding of hearing in the cetaceans, and refute conclusively the errors of the past, which should indeed never have been entertained, for one needs only to consider the fact that cetaceans cannot have any useful olfactory sense, for their noses are tight shut except for brief moments of respiration, and their vision is limited not by any defect of their eyes but by the attenuation of light in passing through water. Even in the clearest of tropical seas objects become obscure at a comparatively short distance, and as distance increases they are soon lost to sight in the characteristic blue-green haze – there is no field of view like that obtaining in clear air where one may see a landscape extending for many miles. Practically no light penetrates from the surface beyond a depth of only 600 feet where it is nearly dark; similarly, even if close to the surface an object separated horizontally from the eye will be invisible at that distance – and in practice at the much shorter distance of approximately 100 feet.

Considering that, like all mammals, whales do not possess a lateral line system similar to that of fishes, that their sense of taste cannot be expected to provide information of phenomena at a distance, nor can any olfactory sense that they may possess, and that, owing to the nature of water, their sight is limited, we must conclude that their very efficient hearing apparatus is the most important to them of all their special senses.

The importance of hearing to cetaceans is confirmed by the structure of the brain, in which the acoustic centres are comparatively much larger than in most mammals. One of them, the superior olive, is notable for its great development; this centre is considered to be concerned with the faculty of echolocation, which we examine below. It is interesting to note that the superior olive is also prominently developed in the bats, the only other mammals that have a highly evolved echolocating system.

It is not surprising that the unusually complex hearing apparatus in cetaceans has more uses than that of most land mammals; it is used not only for ordinary hearing of sounds in the environment but also, among the odontocetes, in hearing those emitted by the animals themselves in the process of sound echolocation or sonar. Considering first the perception of extraneous sounds, a distinction must be drawn between the general sounds incidental to the environment which can give awareness of phenomena occurring in it, and the sounds produced by other cetaceans that may be used in communication. Research with underwater acoustic apparatus in recent years has shown that the old conception of the silent depths of the sea is mistaken, and that on the contrary the sea is a very noisy place. A cetacean must therefore be subject to a continual bombardment of miscellaneous sounds, some of which may convey useful information, but much of which is irrelevant and so merely background noise. A cetacean is no doubt able, like other mammals including ourselves, to disregard those sounds that carry no useful meaning, not through failure to hear them, for they cannot avoid receiving them, but by the way the information derived from them is dealt with in the brain, a sort of voluntary or deliberate cortical deafness. Thus it could be important for a cetacean to be aware of the sound of breakers on a shore, the disturbance produced by a whale catcher vessel, the sounds emitted by a shoal of possible prey, and so on, but irrelevant to attend to the sound of waves at the surface, the snapping of pistol-shrimps or the thousand other noises that bring no useful information, just as we are unaware of the ticking of a clock or other background noises of no importance to us.

The great modern advances in electronic technology have given the cetologists, and for that matter all zoologists, a new method for exploring the environment and lives of animals, including the cetaceans. For the cetologists the new researches grew out of the development of underwater listening and detecting devices particularly directed against submarines during the Second World War. The anti-submariners, most of whom were not biologists, were at first greatly puzzled by what they heard, for they found the seas were full of noises not only from the engines of ships but much more from the sounds emitted by fishes, many crustaceans and other invertebrates, and by cetaceans, some of which were so unexpected in character that the listeners found it hard to believe they were natural sounds and not man-made artefacts.

The users of ultrasonic ranging and direction finding apparatus – ASDIC – soon found that cetaceans could hear, or at least perceive, the ultrasonic emissions, and evidently did not like what they heard, for they invariably fled from them. Sometimes in its retreat a cetacean would release a cloud of bubbles that reflected the Asdic beam and so acted as a screen behind which the animal disappeared, a ploy quickly adopted by submarines when those on board suspected they were being scanned by such a device. They, too, released a stream of air which formed a cloud of bubbles that screened the submarines from observation, causing loss of Asdic contact. Furthermore, the Asdic listeners found that the cetaceans themselves gave out ultrasonic sounds, and measured the range of wavelengths or frequencies that cetaceans use, but such was the secrecy surrounding these wartime operations that knowledge of the cetacean sounds was for long kept from cetologists, with the exception of a handful to whom unauthorized disclosures were made under pledges of confidentiality.

Another great advance in the study of cetacean hearing and sound emission came with the establishment of the great marine aquaria, or oceanariums as they are grandiloquently called, in which the smaller cetaceans could be kept in captivity. As it turned out, the smaller Cetacea proved to be remarkably docile in captivity so that a large range of experiment and observation could be carried out on them at close quarters, using all the latest apparatus provided by the rapid development of electronics. This fascinating work attracted many cetologists so that an enormous amount of information has been collected; indeed, this work of the cetologists, together with similar work on other animals by many zoologists, has become almost a scientific discipline in its own right, that of Bio-acoustics.

The cetologists have not, however, confined their attention to cetaceans in captivity, but have returned to the open seas with the knowledge gained in the aquaria, and have successfully overcome the many difficulties met with in studying and recording the voices of free-living Cetacea in the wild. All this has produced an immense body of published information, and much discussion about the interpretation to be put upon the results obtained, but there is now general agreement about the broad principles involved, though much detail has yet to be resolved. The volume of published results is so great that no more than a summary of the subject can be attempted here.

Although so much is known about the auditory apparatus by which cetaceans hear sounds, very much less is known about the way they produce them. As already mentioned the cetaceans are not provided with vocal cords or folds similar to those found in most land mammals, but they nevertheless possess complex nasal and other air passages in which a body of air can be thrown into vibration. Cetologists are not agreed about whether the sounds emitted by cetaceans are produced in the larynx or in some part or parts of the nasal passages farther from the windpipe. The larynx, which consists of cartilages and projects into the back end of the nasal passages, is provided with muscles, and appears to be suitable for producing sounds. On the other hand there are good grounds for thinking that sounds can also be produced by muscular action on the walls of other parts of the nasal passages; probably both methods are used, perhaps even simultaneously as we shall see shortly.

As cetacean vocal noises are generally made without discharging a stream of air, unlike those of land mammals which usually emit a flow of air through the open mouth, any movement of air must take place between adjoining parts of the air passages. This is perfectly feasible, for we can do the same: if we hold the mouth and nostrils closed we can produce sounds by passing air to and fro through the larynx, and even modulate them into strangulated words and sentences that a hearer can understand. As water is better than air as a transmitter of sound, the vocal noises made internally by cetaceans can be picked up loud and clear by others listening even at great distances. Cetaceans thus certainly do not lack efficient means of communication, and, as far as is known, nearly all of them make use of it in the same general way as other mammals, in producing, for example, recognition cries, cries of alarm, threat and so on. In addition the odontocetes, like bats, use a special part of their voices in echolocation for learning details of their environment, an accomplishment that appears to be lacking in the mysticetes.

In comparison with those of the odontocetes the voices of the mysticetes have been little studied. In the first place it is generally much more difficult to get listening devices close to the mysticetes, and further, the animals themselves are much less loquacious than the odontocetes. The voices of most of the mysticetes that have been studied produce a number of groaning, moaning and rumbling sounds, but little is known about the particular circumstances that stimulate them to emit the different sounds. When the animals are swimming in schools it is assumed that some at least of the noises are useful in maintaining contact between the individuals and so in keeping the school together. It is noteworthy that as far as they have been heard the voices of the different species differ markedly, and consequently they may well serve for mutual recognition.

If this suggestion is correct it may help to explain how the individuals of rare species of cetaceans, such as the ziphiids, which may be scattered over vast areas of the oceans, can find each other for breeding, a question that for long puzzled cetologists before cetacean voices had been discovered. The favourable transmission of sound by sea water makes it quite possible that cetaceans may be able to hear each other when they are as much as several hundred miles apart.

The Humpback is the exception among the mysticetes for it can sometimes produce most dramatic noises. In general this species has been found as taciturn as the other mysticetes, but when it is migrating in warmer waters both of the Atlantic and the Pacific, it behaves very differently. At some times and places the Humpback behaves in an extraordinary way which has been designated, probably correctly, as 'play' because it appears to be produced by no stimulus other than an urge to 'let off steam', a play activity that the behaviourists label a self-rewarding activity and is well known in many mammals. The play, if such it is, takes place generally when the Humpbacks are in coastal waters, but perhaps it is only that observers are more plentiful there, so that similar behaviour in the open oceans goes unnoticed. When Humpbacks play they roll and thresh about on or near the surface and make spectacular breaches, jumping clear of the water and falling back with a tremendous splash; compared with the agile breaching of the smaller odontocetes they may be likened to a herd of elephants imitating a ballet dancer. Whatever the cause of these clumsy gambols, Humpbacks at play also give vent to a wide variety of rather high-pitched calls consisting of squeals, whistlings, organ-pipe notes often with a weird hollow echoing timbre, and other noises extraordinary to human ears, that have been called 'songs' by those who mislead the mass audiences of the 'media'. Many intensely interesting, and aesthetically beautiful, recordings have been made of the 'songs' near Bermuda and Hawaii when Humpbacks are on their breeding migrations.

It is of course possible that the 'songs' are part of the self-rewarding play pattern, but there is no reason for supposing that they are songs in the human connotation of the word any more than are the songs of birds or cicadas, or the braying of jackasses. In view of the fact that Humpbacks in other latitudes have been found to differ little from the other mysticetes in their production of sounds, it may well be legitimate to wonder whether the other species of mysticetes, which have been found to be so quiet, may at some times or places also indulge in similar noisy performances – or is it just that the Humpback is the buffoon of the whalebone whales?

A peculiar point was found by the American cetologists Watkins and Schevill when they were making sound recordings from a Black Right whale in the waters off Cape Cod. The animal was feeding at the surface, skimming up small plankton with the tip of the snout and parts of the front baleen plates out of the water. An unusual sound recorded was found to be caused by the flexible baleen plates clacking together under the influence of the stream of water passing between them, a sound that turned out to be merely incidental to feeding and nothing to do with any purpose-made communicatory sounds [209]. These observers were not able to detect any sounds that could be interpreted as being concerned with echolocation. Nor has any definite evidence of echolocation being used been found for any other of the mysticetes, a matter that is not surprising on anatomical grounds as we shall see when examining the phenomenon in the odontocetes.

Apart from echolocation, the voices of some odontocetes are far better known, although those of only a minority of the large number of species have been heard and recorded. The sounds emitted cover a very wide range of frequencies, from 500 to 25,000 cycles per second or more, and produce low grunts and growls through creaking and scraping sounds to squeals, shrieks and high-pitched whistles. Although every species seems to have a fairly extensive repertoire, the noises of some are characteristic for the species, the liquid trilling whistles, of the White whale, for example, would identify the species to anyone familiar with the calls. The calls may be expressive of emotion or mood, or may be more directly communicative, and some interesting experiments have been made with captive Bottle-nosed dolphins in an attempt to discover whether any specific meaning can be attributed to specific sounds. This was done by recording various calls from the animals and then playing them back underwater and observing the reactions, if any, to them [48].

The experiments identified over thirty distinct calls or whistles, as represented by alterations of pitch against time; thus there are rising whistles, falling ones, rise–fall, fall–rise–fall, and many other variations and repetition trains. When certain signals were injected the animals showed no reaction in their activity but responded with their voices, in some cases repeating the signal given them, in others with different ones. Other signals, however, produced an intense reaction with much activity, strong bursts of echolocation, orientation towards the loudspeaker, and approach to and inspection of it. A single rising whistle seems to be used in response to any new stimulus as a search call, whereas a whistle with a falling contour is a distress or uncertainty reaction. A double-humped whistle was almost invariably associated with impatience, annoyance or irritation. A call that produced an immediate and strong reaction was a whistle with a rise–fall–rise contour. When this was played into a tank containing an adult male dolphin and five females, two of them immature, the animals at once responded with upward search calls, the penis of the male immediately erected and he swam towards the loudspeaker and stopped abruptly in front of it. The dolphins then gave another call with a contour that rose, ran level, and rose again, a call believed to be one to attract the attention of others to the animal emitting it. This experiment produced more activity and variety of vocal response than most, and was in strong contrast to its mirror image, with contour fall–rise–fall, which elicited very little physical activity from the animals; some of them did orientate towards the loudspeaker, but the rest of their behaviour was subdued in contrast with the activity in response to the unreversed call. When a signal with a repeated rise and fall giving a three-humped contour was played the reaction was intense; the animals at once turned towards the loudspeaker, approached and examined it closely, at the same time echolocating heavily. They also responded with a large variety of calls with many minor variations on the 'standard' pattern.

These experiments, which are only an example of the many that have been made, show that some of the calls emitted by cetaceans at least sometimes evoke more marked response from the hearers than others. The amount of information conveyed, however, and the possible specificity of calls, is a matter of controversy. Some have even claimed a precision that seems unwarranted; they have sought to show that the average content for each symbol or pattern or call came to 2.17 'bits' [48]. Others have gone so far as to see in the calls a 'language' with at least some of the characteristics of human speech. The more cautious investigators of the subject, on the other hand, receive these ideas with scepticism; they point out that some of the 'specific' calls are given out when the circumstances are not appropriate, and that equally the expected response from dolphins is often absent when specific calls are presented to them. They also point out that we do not know whether all the parts of the signals as recorded on oscillographs are useful for transmitting information, nor even which parts are or are not useful. The whole subject is one in which it is easy to jump to unwarranted conclusions, and in which enthusiasm or even sentimentality can run away with reason, thus preventing an objective scientific analysis.

Whatever the final upshot of these studies may be, it is safe to conclude that the calls of cetaceans express something of the emotional state of the callers, and that some calls at some times do convey some information to the hearers, as is shown by their sometimes vigorous reactions on hearing them. But to go beyond such generalizations is not justified in the present state of our knowledge, though we may expect the progress of research to bring more precise information in the course of time. The properties of the underwater transmission of sound show that it is possible that cetaceans may communicate vocally over long distances, but we have no information about the realization of such a possibility, or about what such calls may mean to the receiver.

It is important to remember that although a hearer may react to sounds emitted by another cetacean it does not follow that any intention to convey information was held by the sender. Furthermore, any intention to send information must imply a knowledge or expectation of the probable reaction of the receiver. None of these things have been proven, nor does it seem probable that they will be; even an alarm cry, although it may incidentally warn other animals of danger, seems to be relevant only to its producer – does a rabbit squealing in a snare when it sees the trapper approach to knock it over the head expect help or rescue from other rabbits, or intend to warn others of danger? It is inconceivable, and great care must be used to avoid reading more than is justified into what is observed.

The hallucinations and fantasies of the enthusiasts who think that dolphins have a high level of intelligence and are able to communicate with each other and even with man by means of a well developed language are unsupported by any scientific evidence. It is unfortunate that these fallacious ideas have received much publicity and have been widely accepted by an uncritical public lacking the information and means of assessing their value.

An interesting critical experiment has recently been reported [5] in which a pair of Bottle-nosed dolphins separated from each other in a large pool with a visually opaque but acoustically transparent barrier were alleged to communicate by a 'language' in order to obtain a food reward. One dolphin could see two alternative light signals; the other could not see them but could hear the first, and had to press one of two plates as indicated by the appropriate light on being 'told', to obtain a food reward for both animals. The dolphins managed to get a reward on almost every trial, and they were alleged to be communicating by talking to each other in some sort of language. Two animal psychologists, R. Boakes and I. Gaertner, at Sussex University, realized that the results could be explained by well known principles of psychology derived from work on rats and pigeons, including the phenomena known in psychological jargon as 'autoshaping' and 'negative automaintenance'. Neither rats nor pigeons are credited with any linguistic abilities. Boakes and Gaertner repeated the experiment, but used pigeons instead of dolphins, and demolished the fanciful notion of 'language' when the pigeons learned the task well enough to get a reward in 90 per cent of the trials.

The American cetologists Schevill and Watkins have published records of the voices of some eighteen species of cetaceans in the wild, including those of a school of about ninety Pilot whales feeding on squids in Trinity Bay, Newfoundland [179]. The Pilot whales are hunted commercially there, but one Sunday morning in July 'when the whalers were at church' the cetologists were able to take advantage of the Sabbath calm to make their recording. The whales certainly sound excited, making all kinds of whistles, squeals and grunts – they remind the human listener of half a dozen packs of hounds in full cry. One would like to know what effect the uproar has upon the squids; one might well expect it to frighten every squid within earshot into instant flight, though it is hard to suggest where they might go to avoid such a terrifying school of enemies.

In addition to the variety of sounds already mentioned, the odontocetes emit another kind of sound that is heard as a series of short sharp 'clicks' – the brief bursts of sound used in echolocation. Each click consists of a pulse lasting from about one millisecond to twenty or more, and containing vibrations of complex frequencies ranging from four or five to two hundred hertz (kilocycles per second) or more. The repetition rate for the pulses varies according to conditions, but may be some two to five a second in ordinary searching, increased to fifty or more when on target.

The basic principle of echolocation is now widely known, and is similar to that for ships' echosounding. Essentially, a short pulse of high frequency waves is sent out and is reflected from any object solid enough to return them; as the speed of sound is known, the time between transmission and reception can be noted and the distance of the target deduced. This process is done automatically in man-made sonar gear, and doubtless some analogous and automatic process occurs in the cetacean brain. In practice a rapid succession of pulses is transmitted, and if they are directed into a narrow beam they give the direction of the target as well as its distance, for echoes will be returned only when the beam is 'looking at' the object.

The echolocation, or sonar, of cetaceans was discovered by observing the behaviour of captive dolphins, which are able to avoid obstacles or find prey such as small fish in darkness or in water so clouded by suspended silt that sight is impossible. Experiments have shown that for cetaceans sonar is an extremely precise sense, and that it builds up a mental image as complete and detailed as do the eyes of land mammals. The animals 'adjust their sets' in homing on prey or other objects by increasing the pulse recurrence frequency as they approach nearer – the shorter the distance the less the time for the echo to return, and hence a higher pulse rate and increased accuracy are possible. Sometimes the stream of pulses itself makes a note audible to human ears as a creaking sound – rather like the sound produced by the drumming of a woodpecker.

Cetologists are not agreed about the precise source of the ultrasonic vibrations used in sonar; some argue that they are produced in the larynx because a more or less rigid structure with good muscular control is probably more capable of producing the high frequencies required. Purves [160] suggests that in the Sperm whale transfer of air from the lungs to the enlarged chambers of the right nostril gives a reservoir of air for phonation by the larynx at great hydrostatic pressures. On the other hand some cetologists hold that a peculiar rounded structure in the nasal passage of physeterids called the 'museau de singe' is responsible for the production of the sound used in sonar; in other odontocetes a rounded plug connected with the air sacs near the blow-hole is assigned this function. It is possible that both larynx and air passages are used in sound production because odontocetes often make audible sounds at the same time as they are echolocating, though which structure makes audible cries and which sonar clicks has not been definitely decided.

A sonar system to give accurate echolocation must, as already mentioned, be directional as well as ranging, and this requires that the sound pulses be directed into a narrow beam. The American cetologist K. Norris has put forward a theory about how this is achieved, and claims to show what a complex and beautifully arranged anatomical architecture the odontocetes possess and use for echolocation [138]. Dr Norris suggests that the ultrasonic pulses originate in the region of the plug in the nasal passage between the larynx and the blow-hole. In life this structure lies in front of the raised fore part of the skull and above the dished upper surface of the top jaw, so that in effect it lies at the focus of a parabolic reflector. Even if the sound originates in the larynx it could well be reflected from or modulated by the folds in the passage lying at the focus. However that may be, the sound proceeding from this point radiates in all directions, and that passing backwards is reflected from the front of the skull to join the sound waves going directly forward from the source. Similarly the radiation passing obliquely upward is reflected from the surface of the upper air sac, and that passing obliquely downwards is reflected from the upper surface of the upper jaw. All the radiation is thus directed forwards through the melon which is filled with an oil differing from that of the blubber, and which refracts the sound rays, so that it forms an acoustic lens focusing the pulses into a beam directed forwards.

The researches of the American biochemist Professor Carter Litchfield and his colleagues have shown that the lipid composition

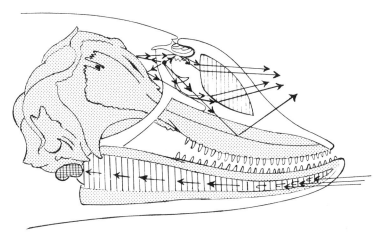

Figure 19. Diagram of the skull and surrounding tissues of a Bottle-nosed dolphin to show the sonar paths according to Norris's theory. The upper arrows trace the reflection of ultrasound from its source by the bone of the skull and its refraction to form a beam by the melon. The lower arrows show the path of the returned echo towards the ear bulla through the oil filling the hollow of the lower jaw.

varies in different parts of the melon fat of some species, and that the character of the melon fat differs in different species. In the Bottle-nosed dolphin the lipid analyses show a division into a central inner melon core surrounded by outer and under melon regions of differing contents. The topography of the melon lipid thus aids in the collimation of the ultrasonic pulses used in echolocation [105]. In general the melon fats consist of more saturated lipids of lower molecular weight than those of the blubber, but differ in different species – in delphinids they consist of triglycerides with a small amount of wax ester, whereas in the phocoenids and monodontids the triglycerides are mixed with little or no wax ester. In the ziphiids on the other hand the blubber fats are almost entirely wax ester and thus differ from those of all other cetaceans. The fact that relative velocity of sound waves through the melon fat differs in different species, being lowest in the delphinids and highest in the physeterids and platanistids, may indicate that there is more than one type of echolocation mechanism within the Odontoceti [104].

Varanasi and her colleagues [207] have studied not only the biochemistry of the lipids but their acoustic properties. They show that relatively small changes in the ratio of wax esters to triglycerides induce

157

substantial alterations in the sound properties of the lipid medium, and that these changes could not be reached through the use of lipids commonly present in most animal tissues. The acoustic lipids in the melon produce lower ultrasonic velocities which enable the necessary sound refraction to occur; they thus form a three-dimensional matrix of ultrasonic velocities that produce a sound lens.

The use of the melon in focusing the ultrasound pulses into a beam is evidently not solely a passive phenomenon, for the shape of the melon can be conspicuously altered by muscular action to make it bulge forward or flatten out. Even in those odontocetes such as the Common porpoise and the Killer, in which the beak is not sharply marked off from the head, the internal structure of the melon is present complete in front of the dish of the fore part of the skull, and functions as in those species in which the beak is more distinct and the melon more prominent. In species without a prominent melon the region is often caused to bulge forward so that it is quite distinct in the living animal though it is less easily recognized after death.

In the preceding paragraphs we have considered the way in which the ultrasonic pulses are transmitted, and it now remains to look at the receiving mechanism. Because the stream of pulses is beamed into a narrow ray straight ahead, reception by the ear passages at each side of the head would be unlikely to be efficient, useful as it is for ordinary sounds. According to Norris's theory the mechanism in the odontocetes deals exactly with this problem. The lower jaw-bone of the odontocetes, unlike that of other mammals, is widely hollow so that a passage with comparatively thin bony walls extends from the tip to the base, gradually increasing in width along its course. The tip lies at or near the end of the snout, whereas the base lies in close relation with the tympanic bulla and middle ear near the jaw joint. The hollow jaw-bones are filled with lipids similar to those of the melon and differing from those of the blubber. Norris suggests, therefore, and some others agree, that the oil-filled jaw-bones act as pathways for conducting the returned echo from the front of the snout to the ear, thus giving the possibility for differential reception and orienting on to the beam.

The whole of the odontocete head in front of the cranium is thus, according to this theory, a complicated transmitter-receiver for pulses of ultrasound. The fore part of the skull is a reflector dish, in front of which lies an acoustic lens that either refracts the sound into a beam or by differential impedance fills the gaps in the polar diagram, or serves both functions, and for reception the oil-filled jaws conduct signals from in front to the middle ear of each side. All this complicated structure shows why echolocation has not been detected in the mysticetes – the shape of the mysticete skull is quite different and has no reflecting dish, there is not a melon, and the structure of the lower jaws differs from that of the odontocetes, being much more solid and without a similar oil-filled internal passage. If the mysticetes do have any echolocation system it must differ fundamentally from that of the odontocetes.

Of all the odontocetes the Sperm whale appears to be the species in which echolocation has reached its peak. Its anatomy alone points towards this: the huge head, which makes about a third of the total body length, contains the enormous reflecting dish of the skull, the 'Neptune's chariot' or 'sledge' of the old whalers, and this holds the huge case filled with spermaceti supported by the junk, though the melon is represented by a small and apparently rudimentary structure. The whole appears to form the largest and most complicated sonar apparatus found in the odontocetes, though it must be remembered that other functions have been ascribed to the spermaceti of the case. However that may be, the Sperm whale seems to have concentrated so much on its sonar that it has given up any other use of its voice; as Schevill and Watson remark, its conversation is remarkably monotonous, for it appears to use sonar clicks for communication as well as for echolocation.

The American cetologists Richard Backus and William Schevill have listened extensively to Sperm whales, and find that their clicks are very loud and can be detected at a distance of several miles [9]. The click rate is rather slow compared with that of dolphins, and each click contains several pulses, the wave form of which is constant in any series of clicks but varies individually so that each whale can recognize its own signals. Some similar signature must be present in all cetacean sonar for without it a school of echolocating dolphins would merely confuse each other. The clicks of Sperm whales are also used in communication, for the animals have never been heard squealing or whistling; Backus and Schevill heard a school of about two dozen Sperm whales idling for an hour at or near the surface 'clicking away in a variable and occasionally almost desultory manner' as the individuals in the school kept contact by 'chatting' to each other. These observers, using the ship's echosounder, also found a Sperm whale about 1,700 feet below the ship in water over 7,000 feet deep about a hundred miles north-east of Cape Hatteras. Then an unexpected thing happened; the whale began answering the 'pings' made by the echo-sounder. The clicks of the whale came at a different rate from the pings of the echo-sounder until the ship was nearly directly above the whale, whereupon it altered its rate so that it gave an answering click every time it heard a ping from the ship. This is taken as showing that the clicks are probably used in communication as well as in echolocation. In echolocating the Sperm whale commonly clicks about twice a second, which indicates that the 1,200 foot stretch of water ahead is all that it can efficiently examine when looking for weak sound scatterers such as squid.

One cannot but wonder what the reaction of squids may be when they are subjected to sonar examination, for it seems improbable that in the course of evolution they have not acquired some sensitivity to it and thus some avoiding reaction. It is well known that cephalopods possess an ink sac and that they discharge the contents as an opaque cloud when they are alarmed or attacked. The cloud of ink is supposed not so much to form a screen behind which the animal can escape, as to provide an apparently dark object that diverts the attention of the pursuer. That is all very well near the surface where light can penetrate, but it cannot work in this way nearly two thousand feet down where no light can reach from above and it is completely dark apart from light emitted by luminous organisms. Can it be that a cloud of cephalopod ink scatters or reflects a sonar pulse to give a false echo, and thus act as part of the squid's escape reaction in the blackness of the great depths? A simple experiment might provide some interesting information.

The acoustic reflector-and-lens theory of echolocation is, however, by no means universally accepted by cetologists as the correct explanation of the sonar system. Schenkkan and Purves have investigated the details of the anatomy of the nasal passages and spermaceti organ in the physeterids [178], and Schenkkan the anatomy and function of the nasal tract in other odontocetes [179]; they suggest differing conclusions. They emphasize that the spermaceti organ is asymmetrical and is not homologous with the melon of other odontocetes, and they show that it cannot function as a hydrostatic organ adjusted by heating and cooling as postulated by Clarke [37]. Its main function is to receive in the sinuses of the right nasal passage the air which is forced from the lungs into them by external pressure during a deep dive; it also aids in expelling the residual air while the whale is 'having his spoutings out' at the surface after a dive. Furthermore, the air in these sinuses expands when the whale starts to return to the surface and hydrostatically aids the ascent. The 'museau de singe' acts as a valve for regulating the passage of air in the sinuses, and has nothing to do with the production of sound. The echolocating clicks are produced by the larynx which at depth draws air from the sinuses for the purpose, so that the small volume of compressed air is recycled between larynx and sinuses.

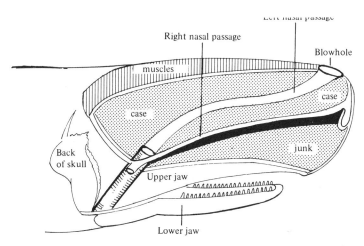

Figure 20. Diagram of the head of a Sperm whale showing the position of the functional left nasal passage leading to the blow-hole, the blind right passage with terminal air sacs, the case and the junk.

These authors doubt the correctness of the reflector-and-lens theory, although they agree that in most odontocetes the whistles and other communicatory sounds may be produced by the accessory air sacs connected with the nasal passages. They hold that all echolocation clicks must be made by the larynx, which is far more suitable for their production than the soft tissues of the nasal passages with less precise muscular control. They point out that the front of the skull cannot act as a parabolic reflector because no cetacean skull resembles a parabolic mirror, 'which is a very precise piece of engineering'. That is true, but the argument is weakened by the fact that some forms of man-made reflector used in radar may be modified from the true mathematical parabola in order to fill gaps in the polar diagram. As for the acoustic

161

lens of the melon, Schenkkan and Purves cast doubt upon the possibility that the oil-filled tissues of the melon act as a lens, mainly because no man-made instrument resembling an acoustic lens has yet been invented. This statement seems to carry less weight in the light of the experiments of Varanasi and her colleagues [207] already mentioned. Furthermore, has any enterprising inventor ever tried to make an acoustic lens?

On the other hand Schenkkan showed that in the Common porpoise, the Common dolphin and other odontocetes the bone of the maxillae and premaxillae of the rostrum is not of uniform density but has strands

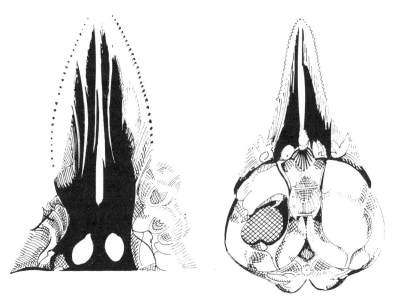

Figure 21. Drawings from X-ray photographs of the skulls of an adult (left) and a newborn (right) Common porpoise to show the areas of denser bone that, according to Schenkkan and Purves, conduct ultrasound forwards.

of dense bone running fore-and-aft in the less dense cancellous surrounding bone. The dense strands are conspicuous in X-ray photographs of the skull, and sound transmission experiments show that maxima of intensity are found at the places where the dense areas in the maxillary bones reach the margin of the rostrum. He suggests further that the dense strands serve for conducting and beaming ultrasonic sound clicks, and may themselves be produced as a reaction to the high intensity of the sound emissions. His sound transmission experiments give convincing polar diagrams of the sound field produced from an odontocete skull provided with an ultrasonic source.

The two theories of the manner of production and transmission of ultrasound for echolocation appear to be mutually exclusive, yet both are supported by much experimental work. We can only await the results of further research to determine their relative merits, and whether they are in fact incompatible in part or as a whole.

Chapter 8

Behaviour and social relations

The Cetacea are essentially gregarious animals, a matter of such common observation that for centuries the collective noun 'school' has been applied to large or small groups of them, a name they share with the fishes. The old whalers similarly used the word 'pod' for a small school of Sperm whales, often a group of females and young males accompanied by a master bull. Schooling is particularly characteristic among many of the odontocetes, schools of some dolphins habitually contain hundreds and sometimes thousands of animals. Schools of Sperm whales are generally much smaller, though they may consist of fifty or more animals, generally mostly females. The mysticetes, on the other hand, rarely form schools of more than half a dozen animals; although many whales may be present on a feeding ground, associations close enough to be called schools generally consist of not more than four or five whales, whether rorquals or Right whales.

In those kinds of mammals that are gregarious the origin of the habit of flocking has been attributed to various environmental and other causes during the course of evolution, though it is now genetically fixed. In cetaceans the schooling habit may well be due to the monotonous character of the environment, for the upper waters of the oceans do not show any topographical differences so that it is practically featureless away from the neighbourhood of the coastlines. For the purpose of reproduction it is essential for animals to be able to find each other at the appropriate season, and the schooling habit obviously facilitates this. In addition, many cetaceans feed on animals that themselves occur in shoals, such as fishes, squids and some Crustacea, or in dense planktonic swarms such as copepods or krill, so that the search for food necessarily brings individual cetaceans together.

In a featureless habitat the presence of other individuals must in itself be an important part of the environment for cetaceans, and the limited range of visibility underwater may also tend to encourage schooling by keeping the animals within sight of each other. If a member of a school loses visual contact with its fellows the use of the voice which, as we have seen, travels underwater for long distances at a speed greater than that of sound in air – about 4,800 feet per second as against about 1,100 feet – will enable visual contact to be regained without delay if desired. Cetaceans, either in schools or as isolated individuals, are probably acoustically aware of the presence and relative positions of others in a very large area of the ocean surrounding them.

Many land mammals have a well marked habit of territoriality by which a limited area becomes more or less private property from which other members of the species are excluded permanently or temporarily. In many small mammals it centres upon a nest or burrow surrounded by a home range used in foraging, and in large mammals it often concerns a temporary area occupied by a proprietor male during the breeding season. Such territories in land mammals are easily defined by topo-graphical features, and are often further defined by marking boundary positions with scent secreted by special glands of the possessor. No such territorial boundaries are available to cetaceans, so they tend to aggre-gate in schools rather than to be widely dispersed as lone rangers, though there are exceptions. It is possible that the area in the neighbourhood of females may represent a sort of territory to males, so that a dominant male may take possession of such a female or school of them and drive other bulls from the vicinity. It has been suggested that this may be a feature of the behaviour of the mysticetes, though no proof that it occurs has been offered. On the other hand the Sperm whale seems to be polygamous, so that a pod of this species consists of a group of females and young males with a proprietor bull which presumably keeps them to himself and drives away any other fully adult male that might try to dispossess him. Support for this idea is claimed by the common occurrence of single adult bulls, or small groups of them, in waters far north or south of the usual range of the species where females are only rarely found. Such bulls are supposed to be aged individuals that have been driven out of the breeding schools by younger and more vigorous animals, though examination of those caught by whalers in these far regions shows that the animals are fully potent sexually and give no sign of feebleness due to age or failing strength.

It seems, however, that there is not an invariable pattern of behaviour in this species, for sometimes pods are found in which there is more than one adult bull, and other pods containing only bulls may cruise in the neighbourhood. Yet fights between rival bulls have been seen, in which the contestants butt each other with the head and try to bite each other, with resulting broken jaws and shattered teeth. Similarly the behaviour of a bull when attacked by whalers was unpredictable; generally the reaction of a harpooned Sperm whale was to flee, often diving to a great depth or swimming away at high speed to windward. Occasionally a bull was harpooned and lanced to death with scarcely a struggle, the animal lying inert while it was killed; less rarely an attacked bull fought back, staving in the whale boat with head or flukes, or seizing it in its jaws and even biting it in half. Had all Sperm whales been equally aggressive the old sailing ship whaling industry might have had a different course.

The most detailed study of the schooling habits of the Sperm whale has, however, been made by the Japanese cetologist Ohsumi [139]. He concludes that Sperm whales are matriarchal in social organization. He calls the fundamental school the 'nursery school', which consists of mature females, with sucklings and immature females and males. Young bulls approaching sexual maturity lose their association with their nursery schools, and make rather looser 'bachelor schools'. 'Harem schools' are only temporary, and are formed by a dominant mature bull taking possession of a nursery school during the breeding season when he serves an average of fourteen cows. Ohsumi considers that mature bulls contend for the possession of harems, and that the scars found on large bulls may be the result of such struggles. At other times mature bulls live separated from the females in loosely organized schools that break up in the breeding season when the bulls scatter widely to take possession of harems. In addition to the bachelor schools he found 'juvenile schools' of mixed sexes of immature whales. Finally, the larger of the old bulls outside the breeding season are generally lone rangers and do not form schools.

The largest school of Sperm whales seen by Ohsumi numbered 120, but as the average size of schools in temperate waters was about twenty he considers that the large schools are formed by the temporary aggregation of several smaller ones. The size of schools is generally small in colder waters. The social tightness of the nursery schools is very close, and the whales live in the same school for many years. Ohsumi's observations agree fairly closely with those of Best made on the west coast of South Africa [19].

The social behaviour of the smaller cetaceans has been observed extensively in several species that have been kept in captivity, especially the Bottle-nosed dolphin, *Tursiops truncatus*. One valuable study on dolphin behaviour was made over a period of four years on a group of twelve Bottle-nosed dolphins in captivity at Marine Studios, Florida [200]. The group consisted of five males and eight females; no new

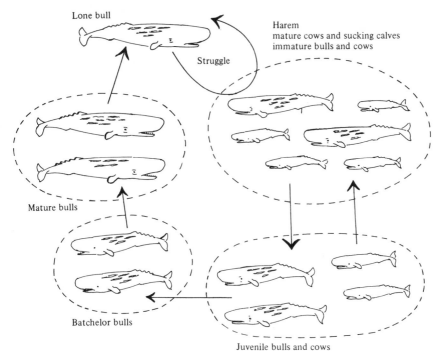

Figure 22. The social structure of Sperm whale schools according to the observations of Ohsumi.

animals were introduced during the course of the observations, the increase on the original number of one male and four females captured as adults being due solely to the birth of young. The group was thus considered to be more 'natural' than other captive groups into which new arrivals were introduced from time to time, with a consequent disturbance of the social structure and necessity for adjustment of the relationship between the newcomers and the members of the existing group, in which social relationships had already been established.

167

In the Marine Studios group there was a definite order of dominance or 'peck order', though it was less pronounced than that found in some land mammals – and animals of other kinds such as birds or fishes. The original male, the largest dolphin of the group, dominated all the others. For most of the time he swam alone, keeping aloof from the others, and seldom became aggressive towards them. Aggressive behaviour sometimes occurred if another tried to snatch food that he was about to take or if a younger male tried to copulate with a female in which he was interested at the time. He signalled his aggression by clapping his jaws loudly, and then chasing and nipping the offender with his teeth and slapping it with strokes of his tail. The victim fled, carrying tooth marks, scratches and bruises on various parts of the body. When the females came into breeding condition the dominant male swam with one or another of the females for several weeks during which she became pregnant. The female concerned submissively swam with him, and at once returned to his side when he called her if she left him for a few moments to feed. She responded favourably to his sexual behaviour and advances, and returned to her normal association with the other females only when he lost interest in her after her oestrus had passed off.

Each female took great care of her offspring and devoted constant surveillance to it for about eighteen months until it was weaned. When her young was born the mother turned on her side to let the infant suckle until it was proficient and could suck when she was in the normal position. At first she kept the infant close to her side, retrieving it at once if it strayed, and keeping it from close contact with the other dolphins or any strange object introduced into the tank. Even after it was weaned the young dolphin might return to rest under its mother, though it spent much time in company with other young ones; she continued to shelter it until she again became pregnant, whereupon she lost interest and pushed it away. When the young were still infants their mothers herded them away from other dolphins whether adult or young ones, and if the infants repeatedly left them they chastised them by swimming on their backs, lifting the infants with the flippers, and holding them in the air above the surface for some moments. The infants wriggled vigorously and 'after a minute of this treatment, the baby was seen to swim quietly by the mother's side for a time, making no further efforts to get away'. The mothers, however, did allow their young to spend long periods in the company of a female that had not borne young and acted as a sort of nursemaid from whose care the young were not herded away.

The second dolphin in the order of dominance was an adult female that had borne several young. She often herded the other dolphins into a tight school, particularly when any new object or person entered the tank, and she took the lead in investigating the new object by echolocation and sight. When she had overcome her suspicions the others also approached, and when they were all satisfied the school broke up. This female was also the leader in the feeding 'shows' and was the first to jump out of the water to take fish from the showman. She sometimes became aggressive and clapped her jaws in the manner of the male; the others reacted with whistling and rapid swimming, and kept away from her until her anger passed.

The rest of the dolphins formed three groups; the other older females, the young males, and the infants, the first, with the dominant female, forming the centre of activity in the tank. They varied considerably in temperament, some being quiet and less inclined to social contacts, others being more active and aggressive. The young males as a group were the mischiefmakers, often interfering with the activities of the others, and only desisting when threatened by the dominant female. Although they were not sexually mature they frequently made sexual advances to the adult females which, however, were not interested and rebuffed their attempts at copulation. The dominant male became aggressive at these activities and attacked them, first clapping his jaws, then chasing the offenders, nipping and biting them with apparent intention to hurt, and buffeting them with heavy blows from his tail.

The frustrated young males, perhaps as a consequence of the vigilance of the dominant male, regularly indulged in homosexual activities, rubbing and stroking each other as in the courtship pattern of adult males and females, erecting their penises and unsuccessfully attempting anal copulation. They also stimulated their erect penises by rubbing them against objects in the tank or by positioning themselves in the strong inflow water jet into the tank. They often went in for activities that were regarded as 'play' in that they brought no reward in food or sexual satisfaction. They frequently played tricks on the keepers who entered the tank wearing breathing apparatus, tweaking the airhose or interfering with the tools being used for maintenance. They also used as playthings objects accidentally dropped into the tank, manipulating them with the mouth and balancing them on flippers and flukes. The possessor of the toy was pursued by the others, who tried to gain possession of it, and the adults soon joined in the game which sometimes continued for hours. The group of youngest animals played similar games, often using the fish provided as food for their elders, being too young themselves to eat it. In another game one released objects in the inlet jet of water while another stationed itself at a catching point farther within the tank to retrieve them. Sometimes the mother of a youngster playing this game would join in.

Dr Tavolga, who made these observations, remarked that the behaviour of the young males was often destructive to the peace and quiet of the community of animals, but their activities 'may have served to use up excessive amounts of energy that would have been used otherwise in the wild state'. This seems to be an important point, for in the wild the animals would doubtless have been engaged in the endless occupation of searching for food, and the play seen in captivity may merely be the result of boredom. The social behaviour seen in this stable group in captivity thus represents the patterns adopted by the animals in captivity, but does not necessarily give a picture of normal behaviour in the wild. The animals adapt their behaviour in accordance with the surroundings in which they find themselves, just as men in prison adapt to the circumstances imposed upon them in a highly artificial and abnormal environment. How far the inference is justified that behaviour in the wild is similar to that in captivity remains to be decided.

Observations made underwater on wild dolphins have found that schools often swim with the animals segregated into layers, with the dominant animals above, nearest the surface, and the subordinate ones below. It has been suggested that the interface between water and air is the most important feature of the environment for all cetaceans, for their lives depend upon reaching it to breathe, and consequently the upper layer in a school is the best place and is the one therefore occupied by the dominant animals. The subordinate animals have to take inferior positions and give way to the dominant ones.

The social organization of the Bottle-nosed dolphin (*Tursiops*) and the Humpbacked dolphin (*Sousa*) has been studied in wild schools off the south-east coast of South Africa by G. Saayman and C. Tayler [172]. *Tursiops* was often seen in schools of one to five hundred animals, sometimes more, which fed by co-operatively herding pelagic fish in the open sea into shoals, so that they quickly caught all they needed and fed 'to satiety'. Such meals were taken mostly in the early morning and late afternoon, forming a diurnal rhythm of breakfast and supper. At other times the dolphins came close inshore and hunted among reefs and in the surf zone in dispersed groups. In contrast *Sousa* occurred in small schools of ten to twenty, and followed habitual routes along the shore to isolated reefs and rocky feeding grounds where they dispersed widely to hunt. When different groups met they were stimulated to unusually lively social interactions; they 'leaped, performed airborne cartwheels, and chased, rubbed against, mouthed, and struck at each other with the flukes'. These actions resemble the normal sexual or 'courtship' display, and apparently acted also as a 'greetings' ceremony, which may well reinforce social bonds when the members of a school unite.

One common activity in the wild does seem to come within the definition of play – the habit of some species of taking a free ride in the bow wave of a ship, and perhaps of wave-riding when suitable ocean waves are available. This activity appears to bring no material reward, and to be indulged in solely for enjoyment. On the other hand, as we have seen in Chapter 5, the animals may in such circumstances merely find themselves in conditions where it is unnecessary to expend any energy in swimming forward, and consequently merely set their flukes at the correct angle and 'free-wheel' without experiencing any pleasurable or other emotion – no animal except man does work that is not necessary. Similarly the skill shown by captive dolphins in manipulating objects may merely represent the skill normally used in securing active prey trying to escape.

It is far too easy for the observers of dolphins to interpret their behaviour in human terms, particularly behaviour designated 'play'. We know the delightful pleasure to be got from swimming, the sense of freedom of movement and weightlessness in a medium that gives practically complete buoyancy, but the cetaceans know no other environment. Although they seem to enjoy the conditions they have no experience of life in a medium that requires the body to support its own weight, and what to us is an enjoyable change is to them merely the normal humdrum condition of life.

Schools of Killer whales are sometimes referred to as 'packs', thereby suggesting that the individuals in a school co-operate in pursuing their prey, and that their actions are the result of intelligence and foresight of the consequences. No one would suggest that Fin whales co-operate in attacking a shoal of planktonic krill, and there is no reason for thinking that on those apparently rare occasions when Killers attack a large whale there is anything more than a free-for-all. It is quite unnecessary to invoke any concept of 'group feeding'.

The American cetologist Tavolga, who made the careful observations on the captive dolphins at Marine Studios, has given a necessary warning about the interpretation of cetacean behaviour [200].

One of the most deceptive, and therefore one of the most dangerous, pitfalls awaiting the observer of dolphins is reliance on anthropomorphism. The bottlenose dolphin in particular, and many other dolphins in general, have built-in smiles, and exhibit types of behaviour which endear them to the observer. Thus the observer is led to describe the behaviour of the animal in human terms, and ascribe to the animal motives that he cannot be sure are actually there, as he is incapable of seeing inside the mind of the animal to determine its purposes. Some descriptions of dolphin behaviour abound in statements of purpose that can properly be ascribed only to humans. It is sincerely to be hoped that such accounts, most of which are misleading and probably inaccurate, will not gain credence in the literature to the extent that they are believed implicitly by other workers in the field.

This statement by one who spent four years studying the behaviour of captive dolphins is a warning that should be taken to heart by all students of animal behaviour.

Apart from the care given by female cetaceans to the welfare of their young, which was for long exploited by whalers through killing a calf in order the more easily to kill the mother who lingered near it, and the solicitous behaviour seen among captive dolphins, there is another type of care-behaviour shown by the odontocetes. The former, known as nurturant behaviour, is almost universal among the mammals, but the latter, or succorant behaviour, in which help is given to an adult in distress, is extremely rare. The American cetologists M. C. and D. K. Caldwell have classified this kind of behaviour in the odontocetes into three types [30]. The first, which they call 'standing by' – the old whalers called it 'heaving to', 'bringing to' and similar nautical names – is shown by a school remaining near or approaching the neighbourhood of a distressed or wounded companion but without giving any assistance. This occurs under conditions that would be expected to lead to alarm and flight; it is sometimes seen also in land mammals, particularly in gregarious species. The second, 'excitement', consists of approaching a comrade in distress and showing evidence of unusual excitement and of distress in the succorant animals. The succorants may even attempt a rescue of the animal in trouble, and this has been particularly noticed in Sperm whales which have been known to attack whale boats that have harpooned a companion, or to push an injured animal away from the source of danger.

The most remarkable is the third, 'supporting behaviour', in which one or more companions support the distressed animal at the surface. This type of behaviour can only occur among cetaceans, which must come to the surface at comparatively short intervals to breathe; in contrast land mammals do not drown however badly they may be injured. The supporting of a distressed odontocete by its companions must surely be an extension of the maternal behaviour of a mother towards its young at birth. The mother pushes the newborn young up to the surface so that it can draw its first breaths, and is often assisted by her companions who appear to be solicitous to help in this early care of the infant animal. It is evidently a strong instinct, or 'innate behaviour pattern', as instinct is now called for short, that causes a cetacean to help another that appears to have difficulty in breaking surface to breathe. It is extraordinary that this kind of behaviour is sometimes given to creatures other than cetaceans.

Many instances have been recorded in which a human swimmer in distress has been helped by dolphins to reach the surface and avoid being drowned. Some at least of these occurrences are properly authenticated and are not merely the legends of folklore. To a dolphin a man struggling in the sea must appear to be the most helpless and inefficient swimmer, so that the instinct to help a companion in need of succour is easily transferred to the distressed human being. This phenomenon has been cited as confirming the alleged intelligence of cetaceans; but were they intelligent they would not aid man, who for the greater part has been a ruthless destroyer of dolphins for centuries. They should rather pull a human swimmer in difficulties down to death by drowning instead of helping him up to life.

There are no grounds, however, for imputing any use of intelligence in such acts, nor even for supposing that the animals know what they are doing. There is no proof that any intention to achieve a specific end is involved, and the actions can be regarded as no more than instinctive reactions released by the right stimulus of seeing an object not behaving normally for a dolphin. This view is supported by the behaviour of a captive dolphin that spent many hours pushing a dead shark to the surface of the water in its tank, in spite of the fact that the dolphin had itself killed the shark shortly before.

Altruistic or 'epimeletic' behaviour is apparently not universal among the cetaceans, for it was not seen in circumstances where it might have been expected in the Indian fresh-water dolphin, *Platanista*. This extraordinary creature lives in the Ganges–Brahmaputra complex and the Indus, the animals in each river system being regarded as different species or at least subspecies. They have an extremely narrow beak with many slender pointed teeth in the jaws, broad paddle-like flippers, and they are functionally blind. Animals from each river system have been studied in detail by Professor Pilleri and his colleagues of the Institute of Brain Anatomy in the University of Berne, both in the field and in captivity, for they were able to bring some alive to aquarium tanks in the laboratory at Berne; they have published their results in a long series of scientific papers [155].

After six months in captivity one of the animals fell sick and during the next four weeks lost strength, swam more slowly, dived for shorter periods between breaths, lost control of its movements, blundered into the corners of the tank, and suddenly died. It was one of two females in the tank, but the other female took no more notice of it when dead than it had while it was alive and in obvious distress; there was no epimeletic behaviour at any time. As the animals were kept only two in a tank they were not able to show any schooling behaviour but the female of an adult pair was dominant to her male companion, although during the breeding season the male was the aggressor. Their social behaviour consisted of swimming in close formation, and in caressing contacts between their bodies. The younger animals in another tank chased each other and played a little with objects put in the tank, though to a much lesser degree than Bottle-nosed dolphins. Much of their 'play' seems to have been inspecting new objects rather than actively using them. In the wild the species has not been seen in schools; no more than two individuals are generally seen together [94].

The eyes of these animals are very small and deep-sunk, and are 'degenerate' to the extent that they cannot form an image, though they appear to be able to distinguish light from darkness. It is not surprising that the eyes are not functional, for the animals live in water so turbid with suspended matter that sight would not be possible to the best of eyes; in correlation their sonar for echolocation is highly organized. The sonar beam is divided so that part is directed ahead and part obliquely downwards, as a result of the shape and position of the air sinuses and of the large crests of the maxillary bones. When searching for food the animals move the head up and down, presumably to scan the area ahead and locate prey.

These river dolphins have a peculiar habit of swimming on one side and lightly brushing the bottom of the tank with the flipper of the lower side. This is supposed to supplement the information about the surroundings obtained by sonar, and to allow the eye of the upper side to perceive light coming from above, though how much can be expected to penetrate the muddy water is not stated. This may enable them to distinguish night from day, though why they should need to is not clear because they are 'continuous transmitters', that is they use their sonar all the time night and day with occasional breaks of a few seconds. Unlike the dolphins with functional eyes, they cannot see even large opaque objects just above the surface, which cannot be located by sonar. Nevertheless such perception may be of value, for the animals are more active by night than by day. As they live in rivers with strong flow-currents they are compelled to keep swimming throughout the twenty-four hours, and their brief periods of silence, when they shut down their sonar, are short naps of a few seconds' duration, the only sleep that they take. On the whole their behaviour seems to be much less lively than that of the marine dolphins, and they can be aware of each other acoustically only by their sonar, for they have never been heard to emit any other sounds such as the whistles and squeals of the marine species.

Pilleri points out that the constant activity of *Platanista* resembles that of blind cave fishes that live in running streams of water underground – but any fish in a river has to keep station actively if it is not to be swept away by the current, unless it can lie up in a sheltered pool or backwater. The jaws of *Platanista* also show a similarity to those of the gharial, a crocodile with extremely narrow long jaws likewise armed with slender needle-pointed teeth, that lives in the same rivers with dark turbid waters. Some might claim that these resemblances show a convergent evolution in adaptation to similar environmental conditions.

The other species of the family Platanistidae, the Amazon river dolphin *Inia geoffrensis* and the La Plata dolphin *Pontoporia blainvillei* are in general less 'specialized' in that they have functional eyes, and produce audible sounds in addition to trains of sonar clicks [31]. Their flippers, like those of their Asiatic relatives, are broad and fan-shaped and may be similarly used as tactile organs in the turbid waters that the animals inhabit. Although some specimens of *Inia* have been successfully held in captivity their behaviour has been less fully documented than that of *Platanista* studied by Pilleri and his co-workers. Nothing is known about the life of the Chinese river dolphin *Lipotes vexillifer* of the Tung-ting lake – very few specimens of the species have been examined by Western cetologists.

Many mammals and other animals show various degrees of imitative behaviour, which is even sometimes taken advantage of by man for his own purposes. In fattening poultry the process can be hastened by keeping the birds together, for a fowl that has eaten to satiation will continue or recommence feeding if it can see another taking food. Among mammals a familiar example is provided by a warren of rabbits out feeding in the open towards sunset: if one animal is alarmed and bolts for home the others copy it and likewise run for safety though they do not know what it may be from which they flee – and similarly with many gregarious mammals. Among the marine dolphins the instinct to imitate the actions of others is particularly prominent, and is probably correlated with the habit of schooling and with the tendency shared with all mammals to investigate any strange object or phenomenon that does not induce alarm and a flight-reaction. It probably helps to keep schools together and, as in so many species, to remove the animals from threatened danger.

The imitative instinct undoubtedly gives one of the reasons why captive dolphins are so easy to train to do circus tricks. It is reinforced by their ability to adapt themselves quickly to the conditions of captivity, and by their docility towards their keepers once they have got to know them. The imitative instinct greatly eases the trainer's task because once he has trained his first dupes to perform as he wishes, newcomers put into the group soon begin to follow the example of their new companions. This was strikingly shown by a Bottle-nosed dolphin that was confined in a tank containing a Spinner dolphin at Marineland of the Pacific. The Spinner is in the habit of leaping from the water and spinning – making two or three rotations on the long axis of its body – before falling back into the water. The Bottle-nosed dolphin put into the tank with the Spinner had never before seen a Spinner or its spectacular performance. When the Spinner jumped out and made its spin in response to the trainer's signal, almost at once the Bottle-nosed dolphin also leaped out and made a similar but less skilful spin [27]. Why the Spinner should habitually spin but other species, though perfectly able to perform the same acrobatics, do not is quite unknown. The benefit, if any, of its habit to the Spinner, and what stimulus causes it to perform the manoeuvre are likewise unknown; perhaps it is again one of those activities which, not being understood, are simply to be labelled 'play'.

Such imitation of activities entirely unnatural to the imitator are frequent – a female False Killer whale, *Pseudorca crassidens*, in a tank at the same establishment quickly learnt to perform tricks by watching her tank companions. She learned to perform all their display entirely on her own by imitating them, and practically trained herself. Training, as that of all performing animals, is usually done by rewarding the dolphins with food on the successful performance of a trick, starting with simple actions and graduating to more complicated ones; the reward given on the correct completion of any trick reinforces the animal's performance every time it is repeated.

Although the brain of the cetaceans is large and shows a complexity of development of the cerebral cortex, there is no reason for supposing that the cetacean capacity for learning is associated with intelligence. It may well be questioned whether the ability to learn is greater than in any other mammal – one need only recollect what the animals in the Russian circus can learn to do, which far exceeds the simple tricks that exhibition dolphins perform. It is true that dolphins can be trained to recognize various cues, perform certain actions, or discriminate various patterns in order to secure a reward of food – but so can rats, hens and other animals that are not regarded as especially intelligent. In all of them patterns of action can be induced and reinforced until they become automatic, but they are not intelligent.

It is generally known that naval authorities of several nations have toyed with the idea of using trained dolphins in warfare, either to fix explosive mines to the hulls of enemy vessels, to discover and destroy enemy mines, or to help detect the presence of enemy submarine craft. These and similar uses are officially denied, although extensive experimental work is believed to have been done. Another activity, however, is officially admitted: the use of trained dolphins for finding, and fixing retrieval gear to, objects lost on the sea floor, though it is doubtful whether they approach the efficiency for these purposes of man-made apparatus, or a human diver within the limits of his endurance.

The imitative instinct may also be implicated in the strange phenomenon of mass stranding which occurs from time to time in several species of odontocetes. A mass stranding occurs when a school of cetaceans comes ashore and is left high and dry by the ebbing tide, the animals being impelled to self-destructive stranding by causes unknown. Some species are notorious for their liability to this happening, but others suffer from it more rarely. Mass strandings occur both on wide sandy beaches and on rocky shores where the underwater profile is fairly steep-to, so the nature of the beach cannot always be concerned with the phenomenon. Furthermore when kindhearted people have sometimes tried to help the animals and have pulled them back into the sea their efforts have been unavailing, for the dolphins thus rescued immediately return and get stranded again as though determined to immolate themselves.

Mass strandings have been observed with many sorts of dolphins, Pilot whales, and even Sperm whales. The False Killer whale, *Pseudorca crassidens*, presents an interesting example. It is a highly gregarious species that is usually oceanic in habit so that it does not normally come into coastal waters. The first specimen known to science was a subfossil skeleton found in the Lincolnshire fens in 1846; it was described by Sir Richard Owen who thought it was an extinct species. Within twenty years, however, some of the animals out of a school of about a hundred were stranded in Kiel Bay, and since then strandings of large schools containing up to more than 200 animals have occurred at irregular intervals on the shores of all the oceans of the world.

Many things have been suggested as the cause of mass strandings: for the False Killer the unusual environment of shallow water into which it sometimes strays from its normal oceanic habitat has been supposed to have disoriented the animals so that they were unable to find their way back to deeper water, but that does not explain why if rescued they immediately return and strand again. It has also been suggested that when oceanic cetaceans get into shallow water, especially where there is a flat sandy or muddy bottom with only a gentle slope towards the shore, their sonar is ineffective because the ultrasound waves undergo specular reflection so that they are deflected away from the animals instead of returning to them. Further, over a nearly level bottom the animals are unable to find which way the slope goes, so that they drive on until they are stranded. These explanations seem highly improbable for animals that have such acute and accurate sonar perception, nor does it account for their immediate return if they are helped back into the sea.

Others have postulated that the mass strandings of cetaceans may be caused by the presence of a parasitic worm or other infection in the middle or inner ear upsetting the sonar system. This also seems to be very unlikely for it is highly improbable that all the members of a school should be afflicted simultaneously – though it might be that if only a few or even one were thus suffering and so got stranded its cries of alarm could stimulate epimeletic behaviour in the others and bring them to an attempted rescue. No such behaviour, however, has been observed.

A recent explanation has been put forward to account for the stranding of White-sided dolphins in Shetland [210]. Large schools of up to a hundred or more

frequently swim inshore and into the more or less land-locked bays or narrow voes of the Shetlands. . . . Finding themselves in shallow water they often panic and with the terrific speeds they then put up, large quantities of sand or mud are whipped off the bottom until the water becomes a brown muddy mixture entirely blacking out all submarine navigation. The dolphins then did not know where to turn. We used to wade off in gum boots and turn their heads seawards but it was no use, they turned around blindly and beached themselves. We noticed that those who got far enough off shore to swim into clear water nearly always managed to reach the ocean again. Obscurity of the water by sand or mud was undoubtedly the cause of their stranding . . .

Plausible as this may sound it does not explain why they do not echolocate themselves back to open water when muddiness obscures vision; indeed echolocation in dolphins was discovered by observation of their agility in muddy water. If they are disoriented by the turbidity how do they locate the beach to which they at once return if pointed in the right direction to escape?

It seems plausible that some sort of panic may be communicated to all the members of a school, just as sometimes happens with other gregarious species, including man – one has only to think of the behaviour of a crowd when a fire breaks out in a crowded public building. What might cause such panic among cetaceans remains unknown, for not every school that comes close inshore gets stranded, so what the stimulus can be that leads to their self-destruction can only be guessed. It seems improbable that the presence of Killer whales, their only enemy excepting man, or even the perception of a far distant school through hearing their calls, can be the reason. Furthermore, there appears to be no record of Killers being seen in the neighbourhood when a mass stranding happens to take place.

Whatever the cause of spontaneous mass stranding may be, it has long been known by man that some kinds of cetaceans can easily be driven ashore in large numbers for his advantage. This manner of hunting Pilot whales has been practised in the northern Atlantic islands for centuries and has more recently been adopted in other waters. When a school of these whales is sighted near the coast a fleet of boats puts out and gets between it and the open sea, whereupon the boatmen frighten the animals by splashing the water so that they are driven into a bay to be stranded on its shore.

This species is also subject to spontaneous mass strandings, but the record of such an occurrence on the coast of Florida in 1971 does not support the suggestion of mass panic afflicting the animals. On the evening of 19 August three Pilot whales out of a herd of over forty stranded on a sandy beach but rejoined the school when they were pushed back into the water by human aid. At about the same time six more stranded about a kilometre away; they too rejoined the school after being repeatedly helped back into the water. Early the next morning the herd was found stranded in about one metre of water on a sandy beach eighteen kilometres to the south; two had died. Repeated attempts were made by the large crowd of people who had assembled to push them off the beach but the animals returned to the shore each time. Finally the two largest were towed away by ropes round their tails and held by boats about 400 yards from the beach, whereupon the rest of the school joined them and swam slowly to about a mile offshore. In the afternoon a collector for a marine aquarium caught one by a rope round the tail and towed it slowly ashore; the rest of the school grouped tightly together and followed his boat, and continued slowly and deliberately towards the beach, about 3 kilometres from the site of the morning stranding.

The beach here was sandy with artificial stone groins at 100 metre intervals.

Some became lodged on the groins, and the crowd of people which had gathered had a difficult time freeing them. These whales suffered numerous cuts from the sharp rocks. The people continued to push and pull individual whales into deeper water, only to have them turn and quite slowly and deliberately swim back toward the beach. . . . Not even the twisting of pectoral fins by people trying to help them produced any active response. This passive attitude was the most striking feature of their behaviour.

Finally the two largest were again towed away and held offshore by ropes round their tails; the rest then joined them when pushed off the beach, though some returned and had to be pushed off again. During the stranding two more were taken to the aquarium. The school joined up and followed the two large whales, which still trailed their tail ropes, and slowly moved out to open water. A sudden storm then stopped further observation.

Five days later twelve or thirteen of the same whales stranded themselves on a beach about 275 kilometres to the south-east. One of the largest was identified by an abnormally shaped dorsal fin and by rope 'burns' on the tail-stock. Six of them were successfully pushed back into the sea, but the others died on the beach. The observers emphasize that the mass stranding did not involve disoriented panic but a deliberate shoreward movement, and conclude that 'Although the cause of the strandings remains unclear, the motivation for them was apparently not momentary, for it seems to have lasted for at least one week and over a distance of 275 kilometres.'

An extraordinary fishery of this kind is traditional in the Solomon Islands for the Spinner dolphin *Stenella longirostris*, which is used for food but more highly prized for its teeth, used as a kind of native currency [44]. The natives have a method of driving a school ashore from a distance of several miles out at sea: they go out in canoes and when they have manoeuvred so that they are seaward of the school they lean over the side and clash together two stones held underwater in the hands. The sound produced in some way drives the dolphins, not in a mad panic but quite gently and slowly, towards the beach. When the animals get into very shallow water they push their snouts into the sand or mud and stand on their heads waving their tails in the air, whereupon the natives on shore run into the water, seize them by the tails and pull them on to the beach for slaughter. No explanation is known for this peculiar behaviour – one might expect the dolphins to dive under the canoes and escape to the open ocean.

Japanese fishermen make large catches of this and other species of dolphin, including the Striped and Bottle-nosed dolphins and False Killers, along the south-east coast of Japan by drive fishery. The schools come inshore in pursuit of squids, and the fishermen, whose usual take is fish not dolphins, co-operate in driving the animals so that schools numbering nearly 3,000 animals are not rarely taken, and the annual catch reaches a total of about 20,000, which is used primarily for human food [134].

Although spontaneous mass strandings may not always be the result of panic, the existence of these drive fisheries, and of other lesser ones in other parts of the world, often opportunistic rather than regularly carried on, demonstrates the ease with which schools of at least some species of cetaceans can be frightened into a self-destructive panic. Were they aware of the circumstances they could easily escape by diving under the boats of their pursuers and swimming to freedom; people who hold that cetaceans are intelligent and reasoning creatures second only to man in their abilities cannot use this particular aspect of cetacean behaviour as an example of intelligence – a school of whales or dolphins appears to show no more intelligence than a frightened flock of sheep.

A final point emerging from the latest researches on the behaviour of cetaceans is that many of the modern scientific discoveries do no more than confirm what Aristotle said about cetaceans more than 2,000 years ago.

Chapter 9

Parasites, diseases and enemies

One might suppose that cetaceans, living for the most part in the clean waters of the oceans, would be free from many of the ills to which land mammals living on a substratum of dirt are liable. Not a bit of it – they are as full of parasites and diseases as slum-dwellers in a shanty town with no sanitation. They are afflicted with commensals, external parasites on their skin, worms and flukes in their internal organs, tumours, and diseases caused by viruses and bacteria.

Cetaceans are not, however, more infested and infected than other mammals, but so many of them have been scientifically examined that our knowledge of their pathology is fairly wide. Great numbers of carcases of the larger whales have been carefully studied by scientists at the whaling stations, and in pelagic floating factories, each of which carries at least one biologist, so that much information on the subject has been collected. The pathology of the smaller Cetacea has been studied not only at commercial fisheries, largely by the Japanese, but also in the many stranded specimens that come to the notice of cetologists from time to time. Stranded dolphins and porpoises are often dead or moribund before they get washed ashore, so the incidence of pathological conditions may seem more common than it really is throughout the population. But every animal must die sooner or later, and those that escape accidental death succumb to infection or an overload of parasites as the waning powers of advancing age lower their resistance – the normal terminal events for them. If this is indeed a true picture the cetaceans must be an unusually fortunate order of mammals to achieve that rarity among the animals, a 'natural' death. It is thus not surprising that many of those casually washed ashore are those enfeebled by approaching mortality.

Infestation with parasites by no means implies that the host is seriously affected or damaged; parasites and their hosts generally reach a balance whereby both survive, for it is fatal to the parasite to kill its host. In addition to the parasites that draw their nourishment from the tissues of the host, numerous forms use cetaceans merely as a substrate and place no burden on the host's metabolism, causing it little inconvenience.

Various kinds of crustaceans are conspicuous among the last category. Among the mysticetes every adult Humpback carries a growth of barnacles, generally of several different kinds. Some of them are acorn barnacles of large size belonging to the genus *Coronula*; one is rather flattened and reaches a diameter of nearly 3 inches, another, not flat, reaches a diameter and height of shell of over $2\frac{1}{2}$ inches. Both stick on the skin of the whale, particularly that of the head, flippers and flukes, but they appear to produce little damage, although they leave a permanent mark on the skin when detached. Two other kinds, *Tubicinella* and *Xenobalanus*, 'pseudo-stalked barnacles', are much more closely associated with the whale, for they are embedded in the skin into which they appear to burrow by downward growth, the first leaving little more than the tip exposed for feeding on microplankton. None of them obtains nourishment from the whale, but the last two certainly damage the skin by being embedded, though they seem to produce no serious reaction such as inflammation. Species of the other division of barnacles, the stalked or ship barnacles, many of which live attached to floating objects in the sea, are also found on whales; at least one species of the genus *Conchoderma* seldom occurs elsewhere. This barnacle does not attach to the skin of the whale but to the shell of one of the acorn barnacles growing on it. Although these epizootic barnacles are common on the Humpback they are also found less frequently on other large whales, particularly the Sperm and Black Right whales, and sometimes on the smaller cetaceans. The Gray whale, too, invariably carries large numbers of barnacles, *Cryptolepas rhacianecti*, mainly on the head. The planktonic larvae of the barnacles evidently settle more readily at metamorphosis on the skin of slow-moving species than on that of the swifter rorquals and dolphins.

The baleen of the mysticetes is frequently infested with ciliate protozoa and a small copepod, *Balaenophila*, which lives in large quantities on the surface of the baleen plates, but these appear to cause no damage or inconvenience to the host, which is almost certainly unaware of their presence. So too with the diatoms *Cocconeis ceticola*, that form a film on the skin, particularly on that of the Blue whale when it has been for some time in Antarctic waters, though it also occurs elsewhere. The film is yellowish-green in colour and is sometimes so conspicuous on the underside that the whalers of the American shore stations knew the Blue whale as the 'Sulphur-bottom'.

In addition to the epizootics, cetaceans are the hosts of certain crustacean skin parasites which live by feeding on their tissues. The whale lice, amphipod crustaceans of the genus *Cyamus* and others closely related, live on the skin of the large whales, most commonly on the Black Right whale, the Humpback and the Gray whale, though they are found more rarely on some other species. They are strongly flattened from upper to under surface, unlike most amphipods which are flattened from side to side, are about an inch in length, and have strong claws at the ends of the legs which enable them to cling tightly to the host's skin. They usually lurk in sheltered places such as the skin folds under the flipper, the genital groove, or among the knobs on the surface of the Humpback. They swarm in conspicuous numbers among the irregularities of the bonnet on the snout of the Black Right whale and in other similar but smaller excrescences on the head and jaws.

At first sight these peculiar cornified growths appear to be pathological, as though produced in reaction to the irritation caused by the whale lice, but this is not the case. Although they provide a convenient habitat for the lice they are genetic in origin, for they are already present in the unborn young. Their usefulness to the whale, if any, is unknown, but they do give the impression that they may preserve the rest of the skin of the whale from damage by providing a place where the lice can concentrate and feed on an overgrowth of the epidermis that can be spared by the host. Fantastic as this notion may seem it is paralleled elsewhere in nature, as for example in the special growths some plants have evolved apparently solely as food for ants which might perhaps otherwise damage them.

Another external parasitic crustacean that is particularly common on the rorquals but is sometimes also found on other species such as the Bottle-nosed whale, is a lernaeid copepod of the genus *Penella*. Lernaeids are usually parasites on the gills or bodies of fish, but one or more species live on cetaceans. They are much larger than the free-living water flea copepods, and highly modified towards their parasitic life. The eggs hatch in the sea and the larval stages are at first free-living, but the females eventually settle on the skin of a cetacean and burrow inwards into the blubber where they undergo a metamorphosis into a creature more resembling a worm than a crustacean. The body becomes a soft sac filled largely with the ovaries, the head end expands into a number of branching processes penetrating the tissues under the blubber from which nourishment is drawn, and the hind end of the body, with two long egg sacs, projects from the surface of the cetacean's skin so that the eggs are released into the water. The projecting part may be up to a foot long and appears as a pink tassel about the diameter of a lead pencil; when all the eggs are released the parasite dies and is sloughed off, after which the site heals over. *Penella* is so small in comparison with the size of its host that it seems improbable that it causes any serious harm, though it may produce enough irritation for the cetacean to be aware of it. Perhaps it even enjoys the tickling stimulation [187].

Skin infections appear to be rare in wild cetaceans; a fungal infestation of the skin by organisms of the genus *Trichophyton* has been occasionally found on various species including Sperm whales in the Antarctic [118]. Skin wounds, such as those produced by the teeth of companions or the bites of small sharks, heal with remarkable speed in wild cetaceans, but the story is very different for those in captivity. However clear the water of an aquarium may be kept by filtration it is impossible to prevent its swarming with bacteria, some derived from the dolphins' intestinal flora and much from the food given, others from the attendants, spectators, and the general surroundings. The bacteria cannot be eliminated by chlorinating the water because a concentration high enough to do so adversely affects the skin and eyes of the cetaceans. In captivity, then, skin wounds inflicted by companions or during capture quickly become infected and fester, often progressing to fatal septicaemia or pneumonia. Veterinarians attending sick captive dolphins find this cause of death so common that at least one cetologist has proposed that unless more care is taken in capturing and handling the animals 'the policy of keeping cetaceans in captivity should be reviewed' [74]. Apart from the skin lesions inflicted during capture and transport, cetacean skin is quickly damaged by drying and sunburn when the animals are stranded or captured and removed from the water.

Although many cetaceans often come into water of low salinity in estuaries and river mouths, long immersion in fresh water is highly damaging to the skin of most species. In fresh water the skin soon softens and swells to a pulpy state so that it sloughs off and becomes ulcerated; only the fresh-water species can withstand its action.

The rorquals and the Humpback are subject to another form of skin damage, the cause of which has long been a mystery. The damage is inflicted when the animals are in temperate or warm seas during their winter migration, and takes the form of shallow pits scattered over the body surface, penetrating an inch or so into the blubber. They are oval in shape, and look as though a large spoon with a sharp edge had been used to take a scoop out of the blubber, cutting through the skin to leave clean sharp edges. In the Antarctic the whales arrive with open wounds, which gradually heal over to leave slightly depressed surface scars with lighter pigmentation than the surrounding skin. The older the scars the less conspicuous they become, so that the differing prominence of each successive series roughly represents a winter migration, though it is not possible to make a definite count. Sometimes the wounds are incomplete, that is the chunk of blubber has not been removed but remains attached by a flap at one end. Similar wounds have been found on smaller cetaceans in warm seas, and also on several kinds of large fishes such as tunny, tuna, swordfish, opah and others.

When these wounds were first studied in the 1920s their cause was unknown, for although they looked as though they might be produced by the bite of some animal, no fish was known that could cut out such a neat scoop-shaped bit. The first conclusion reached was that they were caused by some bacterial infection. This very improbable explanation was superseded by a theory that large lampreys were the culprits; the mouth of a lamprey forms a round or oval sucking disc by which the animal can attach itself to prey such as a fish or cetacean and use the toothed 'tongue' to rasp away the tissues. But it was found that lamprey bites do not make similar scooped-out wounds, nor would a lamprey bite explain the incomplete wounds. It is now known that lampreys do indeed sometimes attack cetaceans but that the wounds they make are not so large as the scoops nor are they so deep and clean-cut [154].

It was not until nearly fifty years later, in 1971, that the real cause of the wounds was identified as a small tropical shark less than two feet long, *Isistius brasiliensis*, which in spite of its name is found throughout the warmer parts of the Atlantic and Pacific Oceans [91]. The lips of this shark, unlike those of others, are fleshy, so that the animal can protrude them against a smooth surface and form a vacuum cup with them by raising the tongue. The shark clings to a passing cetacean or large fish by this sucker action, and scoops or gouges out a lump of flesh or blubber with the row of sharp teeth along the lower jaw – the upper teeth are small and hook-shaped. The shark apparently bites facing the tail of its victim and is rotated by the slip-stream when it has dug in its lower teeth so that the wound is completed to cut out the lump. The final proof that *Isistius* is the scooper was provided by finding hemispheroidal plugs of fish in the stomachs of the sharks, along with the remains of squids. When the shark is unsuccessful in completing its bite a partly detached plug is left filling the wound, and the marks of the small upper teeth can be seen at the attached end.

Cetaceans are also occasionally damaged by swordfish, probably more by accident than by deliberate attack, as the fish is also damaged when its sword breaks off and is left in the wound. Such broken bits of bone sometimes cause extensive abscesses which no doubt heal after their contents have discharged. Cetaceans are also, like all animals, liable to injury from various accidents. A number of cetacean skeletons showing healed fractures of the ribs is preserved in various museums. The cause of such fractures is not known, but it has been plausibly suggested that they may result from fighting between males, as cetaceans not only use their teeth in sexual aggression but charge and butt each other with considerable violence.

The long narrow jaw of the Sperm whale seems to be especially prone to injury; many examples have been recorded in which the lower jaw has been broken and the fracture has united irregularly, leaving the jaw grossly distorted, in others the broken front end of the jaw has been lost. It is surprising that Sperm whales with such injuries are able to secure their food in adequate quantities, yet the animals appear to be in good condition and well covered with blubber. Some of these injuries are doubtless caused by fighting between males; combats have been seen in which each animal tries to seize the jaw of the opponent, approaching from below and swimming on the back; presumably the one that first snaps his jaws shut inflicts the injury on his rival. It has also been suggested that some of these jaw injuries may be caused by the whale colliding with the bottom, though one would expect the animal's sonar to tell it of the nearness of danger – possibly such warning is ignored in the excitement of the chase and capture of a squid. The jaws of the smaller odontocetes are also sometimes found with healed fractures of unknown origin.

Perhaps one of the most bizarre injuries recorded is that of a Sei whale from which one of the flippers was missing. This appeared not to be a congenital deformity but the result of the limb being torn off, and one can only imagine that the animal may have been attacked when young by a Killer whale, which is the only known marine predator that could have done such damage, especially as Killers are known occasionally to set upon the rorquals. A similar injury has been recorded in a Blue whale in which the right flipper was completely missing, the shoulder being healed over and covered with blubber, yet 'the whale's swimming and control had apparently been unaffected'. Mutilated whales with damaged or missing flukes have also been recorded by Raymond Gilmore of the San Diego Museum of Natural History. He mentions two Humpbacks each lacking one tail fluke, two Blue whales with both flukes mutilated and much reduced in size, and a Gray whale that had lost both flukes. Nothing abnormal had been noticed in the swimming of the Humpbacks or the Blue whales before their capture; but the Gray whale was able to swim only by turning on its side and using the flattened tail stock in place of flukes for up-and-down strokes, presumably by contracting the epaxial and hypaxial muscles of each side alternately.

Many kinds of parasitic worms infest the interior of the body of cetaceans. Some of the cestodes, or tapeworms, of whales are of truly cetacean size; *Priapocephalus grandis* reaches a length of nearly 50 feet, and *Diplogonoporus balaenopterae* with a body width of nearly an inch is of similar length. In addition several smaller species infest the large whales. None of them appears to be host-specific and all can occur in several species, some infesting both mysticetes and odontocetes. Although some species are known from only one host, and have been named after it, such as *Trygonocotyle globicephalae*, it is probable that they also infest other species of cetacean. In addition to the adult worms, their encysted larval forms, or bladder worms, are found in the blubber of some cetaceans, though it is not known whether this is a normal part of their life history, for they would be expected to occur rather in the food of the cetaceans [114]. Multiple infestations with two or more species of worm are quite common.

A large number of different species of nematodes, or round worms, infests various parts of the cetacean body, primarily the stomach and intestine. They are often found in great abundance in the stomachs of the smaller odontocetes, and as much as a hundredweight of them has been recorded from that of a Sperm whale. What deleterious effect, if any, they may have on their hosts is not known, though similar heavy worm-burdens in domestic animals are certainly harmful, causing unthriftiness and even death. Nematodes of the genus *Crassicauda* infest the kidneys of cetaceans. The cetacean kidney consists of a large number – up to 5,000 or more in some species – of little kidneys or renculi, like an enormous bunch of grapes pressed closely together, with their drainage tubes joining to form ever larger ducts ending in a single ureter that drains the urine into the bladder. *Crassicauda* is a very long round worm that fixes by its head in the calyx of a renculus, from which the body extends through the ducts to the ureter and even through the bladder into the urethra and penis of the males. A heavy infestation of these worms must almost certainly affect the host adversely, although a slight one appears to be of little consequence.

Flukes, or flatworms, of several kinds have been found in the intestine of cetaceans, and also in the gall bladder like the familiar liver-flukes of the sheep. If they are present in large numbers they cause inflammation and damage in both sites. Cetaceans also, like many land mammals, suffer from lung worms, both trematode or fluke, and nematode or round worm, but whether the morbidity produced is as serious and damaging as it often is in domestic animals remains to be determined. Sometimes the bronchi carry a heavy infestation of round worms which must certainly hamper respiration.

The irritation caused by parasites sometimes leads to the formation of tumours in which they become encysted, but malignant tumours have also occasionally been found in the larger cetaceans. Infestation with parasites also appears to be the usual cause of the stomach ulcers that are sometimes found. Ambergris, that peculiar intestinal concretion found in some Sperm whales, is sometimes held to be a pathological product, but the manner of its formation remains unknown. If it is indeed pathological, it is peculiar that so many Sperm whales should suffer from whatever disability it is that produces it.

Although tuberculosis has been found in captive cetaceans, and pneumonia is often fatal to stranded or captive cetaceans when water is accidentally aspirated into the lungs, no epidemic or pandemic infectious diseases have ever been observed in the wild. The life habits of cetaceans would be expected to be a safeguard against such happenings; however crowded a school of cetaceans may be, even the largest school is so minute compared with the enormous volume of the oceans that, with its constant movement, a rapid dilution and biodegradation of its sewage is assured, so that epidemics are unlikely to afflict them. It is interesting to note that dolphins from tropical and subtropical seas when transferred to captivity in aquaria of temperate regions are liable to contract respiratory infections as a consequence of breathing cold air to which they are not accustomed.

Finally, the odontocetes often suffer from bad teeth. Apart from occasional anomalies such as pairs of teeth fused together and other irregularities of development, such as are to be expected from time to time in any large sample of a mammalian population, there are frequent records of disease-infected teeth. They generally, but not always, occur in aged animals in which the teeth have undergone much wear, sometimes being worn level with the gums – even the massive teeth of the bull Sperm whale wear away with use to become blunt rounded stumps. A different sort of tooth wear is often found in the Killer whale, where the sides of the teeth may be strongly faceted by contact with the teeth of the opposing jaw. This kind of wear is thought to be caused by the large movement allowed by the structure of the lower jaw joint, particularly in a sideways direction [32]. 'In addition, the force with which the animals often attack their prey apparently may frequently result in a permanent misalignment of the mandibles, bringing about atypical and often pathological tooth wear.'

The apparent frequency of the finding of odontocetes with diseased teeth is probably due to the fact that many of the animals examined by cetologists are, as already mentioned, animals enfeebled by senility and consequently those most often washed ashore. The mysticetes, being toothless, cannot suffer similarly, but their baleen, too, undergoes wear in use, although it is kept to its full functional length by growth from the base. There is, however, the case of the Sei whales with deficient baleen that was referred to in Chapter 3, though the cause of the condition, whether pathological or congenital, is unknown.

Apart from the parasites, pathological states, and physical accidents to which cetaceans are potential victims, there are few predators to harass their lives. The only one apart from man is the Killer whale, a member of their own order. Killers are catholic feeders and prey upon practically anything they can catch, from squids and fishes to penguins, seals and the smaller odontocetes, particularly the Common porpoise. They have the reputation of combining as a school to attack individuals of the rorquals, but this appears to be an uncommon happening. They are alleged to attack by biting the lips of their prey and tearing lumps out of its tongue. It is certainly true that they sometimes feed from the carcases of large whales killed by whalers, and do bite lumps out of the protruding tongue, as well as out of the blubber; they will also similarly attack the living calf of a slaughtered whale, as Dudley described as long ago as 1725. But he adds they 'often kill the young ones, for they will not venture upon an old one, unless much wounded'. Dramatic illustrations of Killers attacking rorquals, such as that painted by Millais, are the products of imagination rather than of observation [124]. One of the best authenticated attacks took place in a bay off the coast of Vancouver Island. Here some Killers attacked a Minke whale, the smallest of the rorquals, that appeared to be in difficulties in shallow water. But they appear to have killed the whale by drowning it, for there was no blood in the water, merely an oily scum. When it was dead the Killers ate only the dorsal fin, the tongue and the flesh of the lower jaw, and in addition had managed to nibble off the skin from the rest of the body, leaving the underlying blubber intact [77]. Killers have also been seen to attack Gray whales accompanied by their calves, and once actually to kill the calf, but the mother escaped with little or no damage. It is thus evident that although Killers eat the smaller odontocetes, they are not habitually a menace to the safety of the larger cetaceans.

As all the world knows, man is by far the greatest predator on the cetaceans, and is now by the aid of modern technology overcropping them, together with many other products of the seas. The Esquimaux hunted whales probably for thousands of years before European civilization reached them, but their numbers were small so that their catches had no adverse effects on the stocks of the species available to them. In Europe, however, the early fishery for whales caused a marked decline in the population of Black Right whales, perhaps never very abundant, and the fishery on the coasts of south-western Europe became extinct in the nineteenth century. Long before then it had declined to a small opportunistic fishery, so that in the sixteenth and early seventeenth centuries new whaling grounds were exploited off the coasts of America and in the Arctic. Great numbers of Black Right whales were taken on the first, and of Greenland Right whales on the second.

The American fishery declined in the first part of the nineteenth century when the offshore and distant-waters pelagic fishery for Sperm and other whales developed, including the fishery for Black Right whales in the southern hemisphere. At the same time, and until the end of the century, European whalers continued the northern fishery for Greenland Right whales. As a result the Greenland and both northern and southern Black Right whales became very scarce and the fisheries were abandoned, as was that for Greenland Right whales in the Bering Sea carried on by American whalers from west coast ports. The Californian fishery for Gray whales suffered a similar fate.

These fisheries were carried on by sailing ships, latterly with auxiliary steam, that lowered open whale boats to catch their quarry with hand-harpoons, but although the populations of Right whales suffered severely, those of the Sperm whale were not so seriously affected because being an oceanic species the Sperm whale population was far larger.

The modern era of whaling started when the Norwegian Svend Foyn invented the whale cannon mounted at the bow of a small steam catcher fitted with accumulators to ease the strain on the massive harpoon line and prevent it breaking. The essence of the invention was not the cannon, for swivel guns firing explosive harpoons had long been in use, but the accumulator mounted in a ship large enough to support the weight of a dead rorqual, and with machinery for inflating the carcase with air so that it would float when brought to the surface. The invention made the huge worldwide populations of rorquals available to the whaling industry, and their over-exploitation soon began.

The new method used catchers based on shore stations to which they towed the dead whales for extraction of the oil and other products. The invention at about the turn of the century of the process of hydrogenating whale oil to produce a solid fat greatly increased the demand for whale oil for use in the manufacture of margarine and soap. As a consequence whaling stations were set up in many parts of the world wherever whales could be found not more than about 100 miles from the coast. In the early years of the twentieth century the whalers invaded the Antarctic, setting up whaling stations on the islands of Antarctica, mostly those south of the Atlantic Ocean, to exploit the immense population of rorquals that had been hitherto undisturbed. This new fishery was so large and profitable that many shore whaling stations in other parts of the world became uneconomic and were abandoned, though not before their fishing had considerably depleted the stocks available to them.

Within twenty years the intense fishing in the Antarctic had so reduced the numbers of some species of whale, notably the Humpback, that attention was concentrated more and more on the large rorquals, the Blue and Fin whales. The catchers of a shore whaling station, however, are limited in their radius of action, so that an enormous population of whales remained untapped in other parts of the southern oceans.

The only way to exploit these whales, living where no land existed on which shore stations could be built, was by using pelagic mobile whaling stations – hence the invention of the pelagic floating factory that is completely independent of land. This again was not an entirely new invention, for floating factories had long been in use in addition to shore stations. They were large ships fitted for processing their whales on board after they had been cut up in the water alongside, but they had to be moored in the sheltered waters of a suitable harbour where fresh water was available for the boilers, and where flensers could work from boats alongside the ship. The method was a modern adaptation of the technique of the old sailing pelagic whalers. Sometimes at the beginning of the Antarctic whaling season, before harbours were free from ice, the floating factories began work moored to the ice edge, and it was only a step from this to fitting a floating factory with a slipway up which dead whales could be hauled to the deck for flensing and butchering in the open sea.

Pelagic whaling had another advantage: it was extra-territorial so that it was free from any regulations imposed by Governments on the work of shore stations, as also from taxation on their production, and from licence fees payable to the owners of the land. The first pelagic whaler started work in 1926, and in the following ten years a fleet of larger floating factory ships, each with its attendant whale catchers, invaded the Antarctic. They were so successful that in the middle 1930s they made such an excessive catch that the market was over-supplied with oil, the price slumped, and by mutual agreement the whaling companies sent no expeditions to the Antarctic for a year, so that the market could absorb the surplus oil. This warning went unheeded, and at the end of the Second World War in 1945 whaling was resumed with increased vigour owing to the heavy demand brought by the world shortage of oil.

It was soon obvious that whaling was being overdone, as was shown by the scarcity of the Blue whales and then of Fin whales, so that many companies stopped whaling and sold their fleets, leaving Antarctic pelagic whaling to the Russians and the Japanese, whose ships also exploit the northern and tropical Pacific Ocean. The decline of the industry came not only through the lack of whales but also through the increased world production of vegetable oils which brought a slackening of demand and lowering of price for whale oil. Russia may be assumed to carry on with pelagic whaling not solely for the economic value of the catch, and Japan places more importance on whale meat than whale oil in the effort to feed its large population denied by Japanese geography an adequate home production of animal protein.

Long before economic considerations brought whaling to a decline and the whale population into serious shortage, it was obvious that the story of over-exploitation in the Arctic was being repeated in the Antarctic. Until the invention of modern pelagic whaling it was possible for the countries concerned to have some control of whaling operations in their territorial waters, and to impose restrictions on the catch to preserve the stocks. But whalers operating on the high seas were beyond jurisdiction so that the Governments of whaling countries became anxious for the future of the industry, although the whaling companies seemed determined to reap as big a harvest as possible in the shortest time, regardless of the consequences.

The industry concentrated on the mysticetes, but the Sperm whale was also taken to a smaller extent. The fishery for it still continues owing to the unique properties of sperm oil for various industrial processes. A new substitute for sperm oil has now been announced – the oil extracted from the jojoba bean, the product of a shrub growing wild in the deserts of California and Arizona. The oil is similar to sperm oil in chemical and physical properties, but whether this desert shrub can be brought into cultivation on a scale sufficient to replace sperm oil with jojoba oil remains to be seen.

The International Whaling Commission evolved from earlier organizations, with the aim of preserving the stocks of whales for commercial exploitation and not from any humanitarian, aesthetic or sentimental motivation – to conserve whales for human use but not for the sake of conservation itself. The advice of the scientific consultants of the Commission was for some years regularly overruled by the commercial interests, but since about 1970 more heed has been given to their opinions. However, it is almost a case of locking the stable door after the horse has gone, but at last complete protection is given to some species, including the Blue whale, the number of others agreed to be taken is greatly reduced, certain areas are closed to whaling, and the duration of the whaling season is curtailed.

Other organizations now supplement the International Commission,

195

such as the American Marine Mammal Commission established under the US Marine Mammal Protection Act of 1972, FAO, and Unesco; in addition various countries now have laws controlling or prohibiting the import of the products of cetaceans. There are, too, certain pressure groups advocating the protection of cetaceans for sentimental reasons, which make propaganda directed at an ill informed public to arouse emotional support, though their efforts are sadly hampered by their 'lunatic fringe'.

In addition to the large whales, some of the smaller cetaceans are in danger of being over-exploited. Many local opportunistic fisheries for dolphins and porpoises have been carried on for centuries with no damage to the stocks, but when primitive savages have access to modern fire-arms, and fit outboard engines to their dugout canoes, the probability of over-exploitation is greatly magnified. The biggest regular fishery for small cetaceans is found in the waters of Japan where large numbers of the animals have been caught for human food for many years. The Japanese continue the fishery but are now scientifically studying it and its effects on the stocks; the Whales Research Institute at Tokyo and other scientific organizations are actively engaged.

Small numbers of porpoises and dolphins have always been taken unintentionally or 'incidentally' in the nets of fishermen by whom they are not deliberately captured; such occasional killings are of no importance biologically, but in recent years the development of the Pacific fishery for Yellow-fin tuna has incidentally killed enormous numbers of dolphins, mostly the Spinner dolphin *Stenella longirostris* and the Spotted dolphin *Stenella attenuata*, with smaller numbers of other species. The tuna, invisible below the surface, associate with the dolphins which by their presence show where the shoals of fish are to be found. The fishermen set their purse seines round the dolphins so that they may catch the tuna which are below and behind them. In this way many unwanted dolphins are drowned or fatally injured by getting entangled in the nets. The reason for the association of the tuna and dolphins is quite unknown, but it may be that both are seeking the same food. A manoeuvre known as 'backing down' has been adopted that allows many of the dolphins to escape before the fish are hauled in, but the accidental mortality remains high. In 1974 at one 'set' a tuna boat caught ninety tons of Yellow-fin tuna, and over 2,000 dolphins, mostly Spinners with some Spotted and a few Rough-toothed porpoises [152].

The estimated kill of dolphins by the tuna boats of all nations was

348,000 in 1972, decreasing to 181,000 in 1975; the kill by US boats alone in 1976 was 84,000 to 112,500. The US Marine Mammal Commission finds that the population of dolphins is declining as a result of this wasteful catch, and is studying means of reducing it by developing new techniques of fishing, and by imposing licences for fishing as well as quotas on the permitted catch, and is seeking international co-operation. Some similar fisheries in the Atlantic are perhaps less wasteful because the dolphins, or some of them, are landed for human food.

Thus the most dangerous predator on the cetaceans is at last beginning to appreciate the damage he has inflicted upon the stocks of animals that can have an important place in his economy, and is taking the first steps towards preserving them for future use. Scientific knowledge of the biology of cetaceans can help the exploiters to an assessment of the maximum sustainable yield that can be taken without damaging the stocks. Such estimates depend on the assumption that the stocks are populations of constant numbers, but as no natural population of any animal is permanently stable, no finality can be reached and constant study – 'monitoring' as it is called – is continually necessary.

The only way to restore depleted stocks to their former numbers is to refrain from disturbing them until they have recovered. Thus Blue, Humpback, Right and Gray whales are now protected from commercial whaling everywhere, as are Fin whales in the southern hemisphere, and Fin and Sei whales in the north Pacific. The effects of these regulations are yet to be seen, but there are indications of what may be expected. The most striking example is the American population of the Gray whale which was so reduced in the nineteenth century that its pursuit became unprofitable and ceased. Numbers recovered until in the 1920s the fishery was resumed for a few years, after which it was again abandoned for the same reason. Since then there has not been any commercial fishery and the population has markedly increased, but now a new danger threatens. The tourist industry of California has developed the 'attraction' of whale-watching along the coast and in the lagoons where the whales congregate for breeding. The harassment and disturbance caused by buzzing boatloads of eager 'rubbernecks' is giving concern to the Marine Mammal Commission – so now the official conservationists are devising means for protecting the whales from the depradations of too enthusiastic whale-lovers.

A list of the living species of Cetacea

Compiled from the best authorities.

Order **Cetacea**

Suborder ODONTOCETI	Toothed whales
Family Platanistidae	River dolphins
Platanista gangetica	Ganges dolphin Susu
gangetica indi	Indus dolphin Susu A subspecies of doubtful validity
Inia geoffrensis	Amazon dolphin Bouto
Lipotes vexillifer	Chinese River dolphin
Pontoporia blainvillei	La Plata dolphin Franciscana
Family Ziphiidae	Beaked whales (Most of the species have no true vernacular name. The validity of some is doubtful.)
Mesoplodon pacificus	
bidens	Sowerby's Beaked whale
europaeus	
layardi	Strap-toothed whale Layard's whale
hectori	
grayi	
stejnegeri	
bowdoini	
mirus	True's Beaked whale
ginkgodens	
carlhubbsi	
densirostis	
Ziphius cavirostris	Cuvier's Beaked whale
Tasmacetus shepherdi	
Berardius arnuxii	
bairdii	
Hyperoodon ampullatus	Bottle-nosed whale
planifrons	Southern Bottle-nosed whale

Family Physeteridae

Physeter catodon	Sperm whale
Kogia breviceps	Pygmy Sperm whale
simus	Dwarf Sperm whale

Family Monodontidae

Monodon monoceros	Narwhal
Delphinapterus leucas	White whale Beluga

Family Delphinidae

Steno bredanensis	Rough-toothed dolphin
Sotalia fluviatilis	Buffeo Tucuxí
Sousa chinensis	Indo-Pacific Humpbacked dolphin
teuszii	Atlantic Humpbacked dolphin
Orcaella brevirostris	Irrawaddy dolphin
Peponocephala electra	
Pseudorca crassidens	False Killer whale
Feresa attenuata	Pygmy Killer whale
Orcinus orca	Killer whale Killer
Globicephala melaena	Pilot whale Blackfish
macrorhynchus	Short-finned Pilot whale Blackfish
Lagenorhynchus albirostris	White-beaked dolphin
acutus	White-sided dolphin
obscurus	Dusky dolphin
cruciger	Hourglass dolphin
australis	Peale's porpoise
obliquidens	Pacific White-sided dolphin
hosei	Fraser's dolphin (=*Lagenodelphis hosei*)
Tursiops truncatus	Bottle-nosed dolphin
Grampus griseus	Risso's dolphin
Stenella longirostris	Spinner dolphin
coeruleoalba	Striped dolphin
attenuata	Spotted dolphin
dubia	⎧ One or more
frontalis	⎨ imperfectly known
plagiodon	⎩ Spotted dolphins
Delphinus delphis	Common dolphin
Cephalorhynchus heavisidii	Heaviside's dolphin
commersonii	Commerson's dolphin
hectori	Hector's dolphin
eutropia	Black dolphin
Lissodelphis borealis	Right Whale dolphin
peronii	Southern Right Whale dolphin

Family Phocoenidae

Phocoena phocoena	Common porpoise (US) Harbor porpoise
spinipinnis	Burmeister's porpoise
dioptrica	Spectacled porpoise
sinus	
Phocoenoides dalli	Dall's Harbour porpoise
Neophocaena phocaenoides	Finless porpoise

Suborder MYSTICETI — Whalebone or Baleen whales

Family Balaenidae — Right whales

Balaena mysticetus	Greenland Right whale Bowhead
Eubalaena glacialis	Right whale Black Right whale Northern Right whale
australis	Right whale Black Right whale Southern Right whale
Caperea marginata	Pygmy Right whale

Family Eschrichtiidae — Gray whale

Eschrichtius robustus	Gray whale (formerly *Rachianectes glaucus*)

Family Balaenopteridae — Rorquals and Humpback

Balaenoptera musculus	Blue whale
physalus	Fin whale
borealis	Sei whale
edeni	Bryde's whale
acutorostrata	Minke whale Piked whale
Megaptera novaeangliae	Humpback

References

1. Allen, J. A. 1881. Preliminary list of works and papers relating to the mammalian orders Cete and Sirenia. *Bull. U.S. Geol. Geograph. Surv. Terr.* **6**, 399.
2. Anon. 1837. The ordinary cetacea or whales. *Mammalia*, Vol. VI. Jardine's Naturalists' Library. Edinburgh.
3. ———. 1975. Oil industry worries about narwhals. *Marine Mammal News.* **1**, (5), 5.
4. ———. 1976. Fishes in shoals swim better in their own slime. *New Sci.* **72**, 385.
5. ———. 1977. Pigeons may not talk like dolphins but they knew all about psychology. *New Sci.* **74**, 134.
6. Aristotle (384–22 BC). On the parts of animals. Trans. A. L. Peck. London and Harvard, 1945.
7. ———. Historia animalium. Trans. D'Arcy W. Thompson. Oxford 1910.
8. Arseniev, V. A. 1961. Lesser rorquals of the Antarctic, *Balaenoptera acutorostrata* Lee. *Rep. Conf. Sea Mamm. 1959 Ichthyol. Comm. USSR Acad. Sci.* **12**, 125.
9. Backus, R. H. and Schevill, W. E. 1966. Physeter clicks. In Norris, K. (ed.) Whales, Dolphins and Porpoises. Berkeley and Los Angeles.
10. Baldridge, A. 1972. Killer whales attack and eat a Gray whale. *J. Mammal.* **53**, 898.
11. Bartholin, T. 1654. Historiarum Anatomicarum Rariorum Centuria I et II. The Hague.
12. ———. 1678. De Unicornu Observationes Novae. Amsterdam.
13. Beale, T. 1839. The Natural History of the Sperm Whale. London.
14. Beddard, F. E. 1900. A Book of Whales. London.
15. Belon, P. 1551. L'Histoire Naturelle des Estranges Poissons Marins avec la vraie peincture & description du Dauphin & de plusiers autres de son espèce. Paris.
16. Bennett, F. D. 1836. Notes on the anatomy of the Sperm whale. *Proc. zool. Soc. Lond.* 1836, 127.

17. ——. 1837. On the natural history of the Spermaceti Whale. *Proc. zool. Soc. Lond.* 1837, 39.

18. ——. 1840. Narrative of a whaling voyage round the globe. London.

19. Best, P. B. 1967–70. The Sperm whale, *Physeter catodon*, of the west coast of South Africa. Parts 1–5, *Investl. Rep. Div. Fish. Un. S. Afr.* **61, 66, 72, 79, 80.**

20. Blumenbach, J. F. 1779–80. Handbuch der Naturgeschichte. Göttingen.

21. Bonnaterre, J. P. 1789. Cetologie. In Encyclopédie Méthodique, Tome 183, Paris.

22. Bonner, W. N. 1968. The Fur seal of South Georgia. *Brit. Antarct. Surv. Sci. Rep.* No. 56.

23. Borlase, W. 1758. The Natural History of Cornwall. Oxford.

24. Bourdelle, E. and Grassé, P-P. 1955. Ordre des Cétacés. In Grassé, P-P Traité de Zoologie, **17,** 1st fasc., 341.

25. Bray, W. 1862. (ed). Diary and correspondence of John Evelyn, FRS. New ed. London.

26. Brisson, M. J. 1756. Regnum Animale. Paris.

27. Brown, D. H., Caldwell, D. K. and Caldwell, M. C., 1966. Observations on the behaviour of wild and captive false killer whales, with notes on associated behaviour of other genera of captive delphinids. *Contr. Sci.* No. 95.

28. Browne, S. G. 1954. Dispersal of Blue and Fin whales. *Discovery Rep.* **26,** 355.

29. Buffon, C. L. de. 1749–1804. Histoire naturelle, générale et particulière. Paris.

30. Caldwell, M. C. and Caldwell, D. K. 1966. Epimeletic (care-giving) behaviour in cetacea. In Norris, K. (ed.) Whales, Dolphins and Porpoises. Berkeley and Los Angeles.

31. ——. 1970. Further studies on audible vocalizations of the Amazon freshwater dolphin *Inia geoffrensis. Contr. Sci.* **187.**

32. Caldwell, D. K. and Brown, D. H. 1964. Tooth wear as a correlate of described feeding behaviour by the Killer whale, with notes on a captive specimen. *Bull. S. Calif. Acad. Sci.* **63,** 128.

33. Camper, P. 1820. Observations anatomiques sur la structure intérieure et le squelette de plusiers espèces de cétacés. Paris.

34. Cheever, H. T. 1850. Ed. W. Scoresby. The Whaleman's Adventures in the Southern Ocean. London.

35. Childe, V. Gordon. 1931. Skara Brae. A Pictish village in Orkney. London.

36. Chittleborough, R. G. 1959. *Balaenoptera brydei* Olsen on the west coast of Australia. *Norsk Hvalfangsttid.* **48,** 61.

37. Clarke, M. R. 1976. Oil heating keeps Sperm whales in the swim. *New Sci.* **70**, 357.

38. Clarke, R., Macleod, N. and Paliza, O. 1976. Cephalopod remains from the stomachs of Sperm whales caught off Peru and Chile. *J. Zool., Lond.* **180**, 477.

39. Cole, F. J. 1944. A History of Comparative Anatomy. London.

40. Cuvier, G. 1817. Le règne animal distribué d'après son organisation. Paris.

41. ———. 1823. Recherches sur les ossemens fossiles. Nouvelle édition. Paris.

42. Cuvier, F. 1836. Article 'Cetacea' in Vol. 1. The Cyclopedia of Anatomy and Physiology. Ed. R. T. Todd. London.

43. ———. 1836. De l'histoire naturelle des Cétacés. Paris.

44. Dawbin, W. H. 1966. Porpoises and porpoise hunting in Malaita. *Austr. Nat. Hist.* **15**, 207.

45. ———. 1966. The seasonal migratory cycle of Humpback whales. In Norris, K. (ed.) Whales, Dolphins and Porpoises. Berkeley and Los Angeles.

46. Daubenton, L. J. M. and Desmarest, A. G. 1789. Quadrupèdes et Cétacés. In Encyclopédie Méthodique, Tome 183. Paris.

47. Dewhurst, H. W. 1834. The Natural History of the Order Cetacea, and the oceanic inhabitants of the Arctic regions. London.

48. Dreher, J. J. 1966. Cetacean communication: small-group experiment. In Norris, K. (ed.) Whales, Dolphins and Porpoises. Berkeley and Los Angeles.

49. Duhamel de Monceau, M. 1782. Traité Générale des Pêches. Paris.

50. Dudley, P. 1725. An Essay on the Natural History of Whales. *Phil. Trans. R. Soc. Lond.* **33**, 256.

51. Egede, H. 1745. A Description of Greenland. Trans. from the Danish edition of 1741. London.

52. Erxleben, J. C. P. 1777. Systema Regni Animalis. Leipzig.

53. Eschricht, D. F. 1866 (1862). On the species of Orca inhabiting the Northern Seas. In Flower, W. H. (ed.) Recent Memoirs on the Cetacea. Ray Society.

54. Essapian, F. S. 1955. Speed-induced skin folds in the Bottle-nosed porpoise *Tursiops truncatus*. *Breviora* No. 43.

55. ———. 1963. Observations on abnormalities of parturition in captive bottle-nosed dolphins, *Tursiops truncatus*, and concurrent behaviour of other porpoises. *J. Mammal.* **44**, 405.

56. Fehring, W. K. and Wells, R. S. 1976. A series of strandings by a single herd of pilot whales on the west coast of Florida, *J. Mammal.* **57**, 191.

57. Flower, W. H. 1866. Recent Memoirs on the Cetacea. Ray Society.

58. ——. 1869. On the osteology of the Cachalot or Sperm whale (*Physeter macrocephala*). *Trans. zool. Soc. Lond.* **6**, 309.

59. ——. 1874. On Risso's Dolphin *Grampus griseus* (Cuv.) *Trans. zool. Soc. Lond.* **8**, 1.

60. ——. 1885. List of Cetacea in the British Museum (Natural History) London.

61. ——. 1898. Essays on Museums. London.

62. Flower, W. H. and Lydekker, R. 1891. An introduction to the Study of Mammals living and extinct. London.

63. Fraser, F. C. 1952. Handbook of R. H. Burn's cetacean dissections. London.

64. ——. 1966. Comments on the Delphinoidea. In Norris, K. (ed.) Whales, Dolphins and Porpoises. Berkeley and Los Angeles.

65. ——. 1974. Report on Cetacea stranded on the British coasts from 1948 to 1966. London.

66. Fraser, F. C. and Purves, P. E. 1960. Hearing in Cetaceans. *Bull. British Mus. (Nat. Hist.) Zoology.* **7**, No. 1.

67. Fujino, K. 1962. Blood types of some species of Antarctic whales. *Amer. Naturalist.* **96**, 205.

68. Gesner, C. 1551–8. Historiae Animalium. Zurich.

69. Gilmore, R. M. 1961. The story of the Gray whale. 2 ed. San Diego.

70. Goode, G. B. 1884–7. ed. Fisheries and Fishing Industries of the United States. Washington.

71. Gosse P. H. 1851. A naturalist's sojourn in Jamaica. London.

72. Gray, J. E. 1866. Catalogue of Seals and Whales in the British Museum. 2 ed. London.

73. Gray, J. 1948. Aspects of the locomotion of whales. *Nature, Lond.* **161**, 199.

74. Greenwood, A. G., Harrison, R. J. and Whitting, H. W. 1974. Functional and pathological aspects of the skin of marine mammals. In Harrison, R. J. ed. Functional Anatomy of Marine Mammals. Vol. 2. London.

75. Gronovius, L. T. 1763. Zoophylacii Gronoviana. Leyden.

76. Hakluyt, R. 1589. The principal Navigations, Voyages, Traffiques and Discoveries of the English Nation. London.

77. Hancock, D. 1965. Killer whales kill and eat a minke whale. *J. Mammal.* **46**, 341.

78. Harmer, S. F. 1914. Report on Cetacea stranded on British Coasts 1913. London. Followed by Nos. **2** (1915) to **10** (1927) and continued by Fraser, F. C. 1934 Nos. **11** (1934) to **14** (1974).

79. ——. 1927. Report on Cetacea stranded on British Coasts. No. **10**, 43. London.

80. Harrison, R. J., Brownell, R. L. Jr and Boice, R. G. 1972. Reproduction and gonadal appearances in some odontocetes. In Harrison, R. J. ed. Functional Anatomy of Marine Mammals. Vol. 1. London.

81. Harrison, R. J. and Thurley, K. W. 1972. Fine structural features of delphinid epidermis. *J. Anat.* 111, 498.

82. Hart, T. J. 1942. Phytoplankton periodicity in Antarctic surface waters. *Discovery Rep.* 21, 261.

83. Hertel, H. 1966. Structure, form and movement. New York.

84. Hill, J. 1752. An History of Animals. London.

85. Hinton, M. A. C. 1925. Reports on Papers left by the Late Major G. E. H. Barrett-Hamilton relating to the Whales of South Georgia. Crown Agents for the Colonies. London.

86. Hoese, H. D. 1971. Dolphins feeding out of water in a salt marsh. *J. Mammal.* 52, 222.

87. Hisokawa, H. and Kamiya, T. 1971. Some observations on the cetacean stomachs, with special considerations on the feeding habits of whales. *Sci. Rep. Whales Res. Inst.* 23, 91.

88. Houck, W. J. 1962. Possible mating of Grey whales on the northern California coast. *Murrelet.* 43, 3.

89. Hubbs, C. L. 1959. Natural History of the Gray whale. Proc. XV Internat. Congress Zool. London.

90. Hunter, J. 1787. Observations on the structure and oeconomy of whales. *Phil. Trans. Roy. Soc. Lond.* 77, 371.

91. Jones, E. C. 1971. *Isistius brasilensis*, a squalid shark, the probable cause of crater wounds on fishes and cetaceans. *Fishery Bull.* 69, 791.

92. Jungklaus, F. 1898. Die Magen der Cetaceen. *Jena Z. Naturw.* 32, 1.

93. Kasuya, T. 1972. Growth and reproduction of *Stenella caeruleoalba* based on the age determination by means of dental growth layers. *Sci. Rep. Whales. Res. Inst.* 24, 57.

94. Kasuya, T. and Aminul Haque, A. K. M. 1972. Some information on distribution and seasonal movement of the Ganges dolphin. *Sci. Rep. Whales Res. Inst.* 24, 109.

95. Kasuya, T., Miyazaki, N. and Dawbin, W. H. 1974. Growth and reproduction of *Stenella attenuata* on the Pacific coast of Japan. *Sci. Rep. Whales Res. Inst.* 26, 157.

96. Kasuya, T. and Rice, D. W. 1970. Notes on baleen plates and on arrangement of parasitic barnacles of Gray whale. *Sci. Rep. Whales Res. Inst.* 22, 39.

97. Kawamura, A. 1974. Food and feeding ecology in the southern Sei whale. *Sci. Rep. Whales Res. Inst.* 26, 25.

98. ——. 1975. A consideration on an available source of energy and its cost

for locomotion in Fin whales with special reference to the seasonal migrations. *Sci. Rep. Whales Res. Inst.* **27**, 61.

99. Kerr, R. 1792. The Animal Kingdom or Zoological System of the celebrated Sir Charles Linnaeus. London.

100. Lacépède, B. G. de la Ville. 1804. Histoire naturelle de Cétacées. Paris.

101. Laws, R. M. 1958. Recent investigations of Fin whale ovaries. *Norsk Hvalfangsttid.* **47**, 225.

102. Lesson, R. P. 1828. Histoire naturelle générale et particulière des mammifères et des oiseaux. Vol. I. Paris.

103. Linné, C. von. 1758. System Naturae. 10 ed. Stockholm.

104. Litchfield, C., Greenberg, A. J., Caldwell, A. K., Caldwell, M. C., Sipos, J. C. and Ashman, R. G. 1975. Comparative lipid patterns in acoustical and non acoustical fatty tissues of dolphins, porpoises and toothed whales. *Comp. Biochem. Physiol.* **50 B**, 591.

105. Litchfield, C., Karol. R. and Greenberg, A. J. 1973. Compositional topography of melon lipids in the Atlantic Bottlenosed dolphin, *Tursiops truncatus*. Implications for echolocation. *Marine Biol.* **23**, 165.

106. Lönnberg, E. 1906. Contributions to the Fauna of South Georgia. 1. Taxonomic and biological notes on vertebrates. *K. Svenska Vetensk. Akad. Handl.* **40**, No. 5.

107. Mackintosh, N. A. 1965. The Stocks of Whales. London.

108. ——. 1966. The distribution of southern Blue and Fin whales. In Norris, K. (ed.) Whales, Dolphins and Porpoises. Berkeley and Los Angeles.

109. ——. 1970. Whales and krill in the twentieth century. In Holdgate, M. (ed.) Antarctic Ecology, Vol. I. London.

110. Mackintosh, N. A. and Wheeler, J. F. G. 1929. Southern Blue and Fin whales. *Discovery Rep.* **1**, 257.

111. Major, J. D. 1673. De Anatome Phocaenae, vel Delphini septentrionalum *Misc. Cur. Medioco-Physica. Acad. Nat. Curio.* **Ann. 3**. (1672) Obs. XX, 25.

112. ——. 1673. De Igne ex Glacie. *Misc. Cur. Medico-Physica Acad. Nat. Curio.* **Ann. 3**. (1672) Obs. CII, 165.

113. Makarov, R. R., Naumov, A. G. and Shevtsov, V. V. 1970. The biology and the distribution of Antarctic krill. In Holdgate, M. (ed.) Antarctic Ecology Vol. I. London.

114. Markowski, S. 1955. Cestodes of whales and dolphins from the Discovery collections. *Discovery Rep.* **27**, 379.

115. Martens, F. 1675. Friderick Martens von Hamburg Spitzbergische oder Groenlandische Reise-Beschreibung gethan im Jahr 1671. Hamburg.

116. Matthews, L. Harrison. 1932. Lobster krill. Anomuran crustacea that are the food of whales. *Discovery Rep.* **5**, 467.

117. ———. 1938. The Humpback whale, *Megaptera nodosa*. *Discovery Rep.* **17**, 7.

118. ———. 1938. The Sperm whale, *Physeter catodon*. *Discovery Rep.* **17**, 95.

119. ———. 1938. Notes on the southern Right whale, *Eubalaena australis*. *Discovery Rep.* **17**, 169.

120. ———. 1938. The Sei whale, *Balaenoptera borealis*. *Discovery Rep.* **17**, 185.

121. McCann, C. 1962. The taxonomic status of the beaked whale *Mesoplodon hectori* (Gray) – Cetacea. *Rec. Dom. Mus. Wellington*. **4**.

122. ———. 1974. Body scarring on cetacea – odontocetes. *Sci. Rep. Whales Res. Inst.* **26**, 145.

123. Merrett, N. R. 1963. Pelagic Gadoid fish in the Antarctic. *Norsk Hvalfangsttid*. **52**, 245.

124. Millais, J. G. 1906. The Mammals of Great Britain and Ireland. Vol. 3. London.

125. Mitchell, E. 1970. Pigmentation pattern evolution in delphinid cetaceans: an essay in adaptive coloration. *Can. J. Zool.* **48**, 717.

126. ———. 1975. Trophic relationships and competition in northwest Atlantic whales. In Burt, M. D. B. (ed.) *Proc. Can. Soc. Zool.* Annual meeting 1974.

127. Miyazaki, N. Kusaka, T. and Nishiwaki, M. 1973. Food of *Stenella caeruleoalba*. *Sci. Rep. Whales Res. Inst.* **25**, 265.

128. Moiseev, P. A. 1970. Some aspects of the commercial use of the krill resources of the Antarctic seas. In Holdgate, M., (ed.) Antarctic Ecology. Vol. I. London.

129. Monceau, Duhamel de. 1782. Traite Générale des Pêches, et Histoire des Poissons qu'elles fournissent. Paris.

130. Monro, A. 1785. The structure and physiology of fishes. Edinburgh.

131. Morejohn, G. V. and Baltz, D. M. 1972. On the reproductive tract of the female Dall porpoise. *J. Mammal.* **53**, 606.

132. Nemoto, T. 1970. The feeding pattern of baleen whales in the ocean. In Steel, J. H. (ed.) Marine Food Chains. Berkeley and Los Angeles.

133. Nemoto, T. and Yoo, K. I. 1970. An amphipod (*Parathemisto gaudichaudii*) as a food of the Antarctic Sei whale. *Sci. Rep. Whales Res. Inst.* **22**, 153.

134. Nishiwaki, M. 1975. Ecological aspects of smaller cetaceans, with emphasis on the Striped dolphin (*Stenella caeruleoalba*). *J. Fish. Res. Bd. Canada.* **32**, 1069.

135. Nishiwaki, M. and Hayashi, K. (1950). Copulation of Humpback whales. *Sci. Rep. Whales Res. Inst.* **3**, 183.

136. Nishiwaki, M. and Kasuya, T. 1970. A Greenland Right whale caught at Osaka Bay. *Sci. Rep. Whales Res. Inst.* **22**, 45.

137. Norman, J. R. and Fraser, F. C. 1948. Giant Fishes, Whales and Dolphins. London.

138. Norris, K. 1964. Some problems in echolocation in cetaceans. In Tavolga, W. N. (ed.) Marine Bio-acoustics. New York.

139. Ohsumi, S. 1971. Some investigations on the school structure of the Sperm whale. *Sci. Rep. Whales Res. Inst.* **23**, 1.

140. Oliver, W. R. B. 1937. *Tasmacetus shepherdi*: a new genus and species of beaked whale from New Zealand. *Proc. zool Soc. Lond.* **B 197**, 371.

141. Omura, H., Ohsumi, S., Nemoto, T., Nasu, K. and Kasuya, T. 1969. Black Right whales in the north Pacific. *Sci. Rep. Whales Res. Inst.* **21**, 1.

142. Owen, R. 1861. Essays and observations on Natural History, Anatomy, Physiology, Psychology and Geology by John Hunter, F.R.S. London

143. ——. 1868. On the anatomy of vertebrates. Vol. III. London.

144. ——. 1869. On some Indian cetacea collected by Walter Elliot, Esq. *Trans. zool. Soc. Lond.* **6**, 17.

145. Packard, A. 1972. Cephalopods and fish: the limits of convergence. *Biol. Revs.* **47**, 241.

146. Palmer, E. and Weddell, G. 1964. The relationship between structure, innervation and function of the skin of the Bottlenose dolphin (*Tursiops truncatus*). *Proc. zool. Soc. Lond.* **143**, 545.

147. Palmer, J. F. 1837. The Works of John Hunter, FRS, with notes. 5 vols. London.

148. Parry, D. A. 1949. The anatomical basis of swimming in whales. *Proc. zool. Soc. Lond.* **119**, 49.

149. ——. 1949. The structure of whale blubber, and a discussion of its thermal properties. *Quart. J. micro. Sci.* **90**, 13.

150. ——. 1949. The swimming of whales and a discussion of Gray's paradox. *J. exp. Biol.* **26**, 24.

151. Payne, M. R. 1977. Growth of a Furseal population. *Phil. Trans. Roy. Soc. Lond.* **B. 279**, 67.

152. Perrin, W. F. and Watkin, W. A. 1975. The Rough-toothed porpoise, *Steno bredanensis*, in the eastern tropical Pacific. *J. Mammal.* **56**, 905.

153. Peterlin, A. 1970. Molecular model of drag reduction by polymer solutes. *Nature, Lond.* **227**, 598.

154. Pike, G. C. 1951. Lamprey marks on whales. *J. Fish. Res. Bd. Canada.* **8**, 275.

155. Pilleri, G. (ed.) 1969–76. Investigations on cetacea. Vols 1–7. Waldau, Berne.

156. Pliny (Holland, P. 1601) The Historie of the world, commonly called

The Natural Historie of C. Plinius Secundus. Translated into English by Philemon Holland, Doctor of Physicke. London.

157. Pontoppidan, E. 1755. The Natural History of Norway. [Trans. from the Danish ed. of 1751] London.

158. Purves, P. E. 1955. The wax-plug in the external auditory meatus of the mysticeti. *Discovery Rep.* **27**, 293.

159. ——. 1963. Locomotion in whales. *Nature, Lond.* **197**, 334.

160. ——. 1966. Anatomy and physiology of the outer and middle ear in cetaceans. In Norris, K. ed. Whales, Dolphins and Porpoises. Berkeley and Los Angeles.

161. Purves, P. E. and Mountford, M. D. 1959. Ear plug laminations in relation to the age composition of a population of Fin whales. *Bull. Brit. Mus. (Nat. Hist.) Zoology* **5**, No. 6.

162. Purchas, S. 1625. Hakluytus Posthumus or Purchas his Pilgrimes contayning a History of the World in Sea Voyages and Lande Travells, by Englishmen and others. London.

163. Rabelais, F. 1653. The Works of Mr Francis Rabelais Doctor in Physic. Containing five books of the Lives and Heroick Deeds and Sayings of Gargantua and his Sonne Pantagruel. [Trans. into English by Sir Thomas Urquhart]. London.

164. Rapp, W. 1837. Die Cetaceen. Stuttgart and Tubingen.

165. Ray, J. 1671. An account of the dissection of a porpus. *Phil. Trans. Roy. Soc. Lond.* **6**, 2220, 2274.

166. Ray, G. Carleton and Schevill, W. 1974. Feeding of a captive Gray whale *Eschrichtius robustus. Marine Fisheries Rev.* **36**, 31.

167. Rayner, G. W. 1940. Whale marking: progress and results to December 1939. *Discovery Rep.* **19**, 245.

168. Rice, D. W. 1961. Sei whales with rudimentary baleen. *Norsk Hvalfangsttid.* 1961, No. 5, 189.

169. Rice, D. W. and Wolman, H. A. 1971. The life history and ecology of the Gray whale (*Eschrichtius robustus*). *Amer. Soc. Mammal., Spec. Publ.* No. 3.

170. Rondelet, G. 1554–5. Universae aquatilium Historiae. Lyons.

171. Ruud, J. T. 1940. The surface structure of the baleen plates as a possible clue to age in whales. *Hvalrådets Skrift.* **23**, 1.

172. Saayman, G. S. and Tayler, C. K. 1973. Social organization of inshore dolphins (*Tursiops aduncus* and *Sousa*) in the Indian Ocean. *J. Mammal.* **54**, 993.

173. Sargent, J. R. 1975. The unique role of lipid wax esters in the marine food chain. *Nat. Env. Res. Council News J.* No. 12, 4.

174. Sayed, Z. El-sayed. 1970. On the productivity of the Southern Ocean. In Holdgate, M. ed. Antarctic Ecology. Vol. 1. London.

175. Scammon, C. M. 1874. The Marine Mammals of the North-western Coast of North America ... with an Account of the American whale-fishery. San Francisco.

176. Scheffer, V. B. 1953. Measurements and stomach contents of eleven delphinids from the northeast Pacific. *Murrelet.* **34**, 27.

177. Schenkkan, E. J. 1973. On the comparative anatomy and functions of the nasal tract in odontocetes (Mammalia, Cetacea) *Bijdrag t.d. Dierkunde.* **43**, 127.

178. Schenkkan, E. J. and Purves, P. E. 1973. The comparative anatomy of the nasal tract and the function of the spermaceti organ in the Physteridae (Mammalia, Cetacea). *Bijdrag. t.d. Dierkunde.* **43**, 93.

179. Schevill, W. E. and Watkins, W. A. 1962. Whale and porpoise Voices. Phonograph disc. Woods Hole, Mass.

180. Schmidt-Nielsen, K. 1972. Locomotion: energy cost of swimming, flying and running. *Science. N.Y.* **177**, 222.

181. Scholander, P. F. 1940. Experimental investigations on the respiratory function in diving mammals and birds. *Hvalrådets Skrift. Oslo* **27**, 1.

182. ——. 1959. Wave-riding dolphins: how do they do it? *Science. N.Y.* **129**, 1085.

183. Scopoli, J. A. 1777. Introductio ad Historiam Naturalem. Prague.

184. Scoresby, W. 1820. An account of the Arctic regions with a History and Description of the Northern Whale-fishery. Edinburgh.

185. ——. 1823. A Journal of a Voyage to the Northern Whale-fishery. Edinburgh.

186. Scoresby-Jackson, R. E. 1861. The life of William Scoresby. London.

187. Scott, M. 1836. Tom Cringle's Log. Paris.

188. Sergeant, D. E. 1962. The biology of the pilot or pothead whale *Globicephala melaena* (Trqil) in Newfoundland waters. *Bull. Fish. Res. Bd. Canada.* **132**, 1.

189. ——. 1976. Marine populations. *New Sci.* **72**, 300.

190. Sergeant, D. E. and Brodie, P. F. 1975. Identity, abundance and present status of populations of White whales, *Delphinapterus leucas* in North America. *J. Fish. Res. Bd. Canada* **32**. 1047.

191. Shaw, G. 1800–26. General Zoology or Systematic Natural History. London.

192. Sibbald, R. 1684. Scotia Illustrata, sive Prodromus Historiae Naturalis. Edinburgh.

193. ——. 1692. Phalainologia Nova: sive Observationes de rarioribus quibusdam Balaenis in Scotiae Littus nuper ejectis. Edinburgh.

194. Slijper, E. J. 1958. Walvissen. Amsterdam.

195. Sokolov, V., Bulina, I. and Rodionov, V. 1969. Interaction of dolphin epidermis with flow boundary layer. *Nature, Lond.* **222**, 267.

196. Southwell, T. 1881. The Seals and Whales of the British Seas. London.
197. Stamp, T. and Stamp, C. 1976. William Scoresby, Arctic Scientist. Whitby.
198. Starbuck, A. 1878. History of the American whale fishery from its earliest inception to the year 1876. *Rep. U.S. Comm. Fish.* 1875–6. Pt. IV, appendix A, Washington.
199. Sund, P. N. 1975. Evidence for feeding during migration and of an early birth of the Californian Gray whale (*Eschrichtius robustus*). *J. Mammal.* **56**, 265.
200. Tavolga, M. C. 1966. Behavior of the Bottlenose Dolphin (*Tursiops truncatus*): Social interactions in a captive colony. In Norris, K. (ed.) Whales, Dolphins and Porpoises. Berkeley and Los Angeles.
201. Topsel, E. 1607. The Historie of Foure-footed Beastes. London.
202. Townsend, C. A. 1935. The distribution of certain whales as shown by log-book records of American whaleships. *Zoologica, N.Y.* **19**, 1.
203. True, F. W. 1904. The Whalebone Whales of the Western North Atlantic. *Smithsonian Contr. Knowledge. XXXIII. Washington.*
204. Tyson, E. 1680. Phocaena, or the Anatomy of a Porpus. London.
205. van Beneden, P. J. 1885. Histoire naturelle des Cétacés des mers d'Europe. *Mem. Acad. Roy. des Sci. Belge.* XLI. [Published as a separate work with same title. 1889. Brussels.].
206. van Beneden, P. J. and Gervais, M. 1880. Ostéographie des cétacés vivants et fossiles. Paris.
207. Varanasi, U., Feldman, H. R. and Maluis, D. C. 1975. Molecular basis for formation of lipid sound lens in echolocating cetaceans. *Nature, Lond.* **225**, 340.
208. Vesalius, A. 1725. Opera omnia Anatomica et Chirurgica. Leyden.
209. Watkins, W. A. and Schevill, W. E. 1976. Right whale feeding and baleen rattle. *J. Mammal.* **57**, 58.
210. Williamson, C. J. 1977. Pilot whales, *New Sci.* **73**, 481.
211. Williamson, G. R. 1975. Minke whales off Brazil. *Sci. Rep. Whales Res. Inst.* **27**, 37.
212. Willughby, F. 1685. Ichthyographia. London.
213. Worm, O. 1655. Museum Wormianum. Leyden.
214. Yablokov, A. V. 1963. On the types of colour of cetacea. *Byull. Mosk. Obshch. Isp. Prir. Otd. Biol.* **68** (6) 27. In Russian with English summary.
215. Zenkovich, B. A. 1970. Whales and plankton in Antarctic waters. In Holdgate, M. (ed.) Antarctic Ecology. Vol. 1. London.
216. Zorgdrager, C. G. 1720. C. G. Zorgdragers Bloeyende Opkomst der Aloude en Hedenaagsche Groenlandsche visschery. Amsterdam.

Index